The Nature of the Beast

The Nature of the Beast

The first scientific evidence on the survival of apemen into modern times

BRYAN SYKES

CORONET

First published in Great Britain in 2015 by Coronet
An imprint of Hodder & Stoughton
An Hachette UK company

1

A CIP catalogue record for this title is available from the British Library

ISBN 9781444791259
Trade Paperback ISBN 9781444791266
Ebook ISBN 9781444791273

Typeset in Minion Pro by Palimpsest Book Production Limited,
Falkirk, Stirlingshire

Printed and bound by Clays Ltd, St Ives plc

Hodder & Stoughton policy is to use papers that are natural, renewable
and recyclable products and made from wood grown in sustainable
forests. The logging and manufacturing processes are expected to
conform to the environmental regulations of the country of origin.

Hodder & Stoughton Ltd
338 Euston Road
London NW1 3BH

www.hodder.co.uk

Contents

PART 1

1

The Big Guy

The events I am about to describe defy any rational explanation; something that as a scientist who believes in the triumph of reason over superstition, I find profoundly disturbing. The events occurred in the western margins of the northern Cascade Mountains about a hundred miles north of Seattle. I was taken there by Lori Simmons, a young woman in her thirties, who has dedicated a large part of her life to carrying on her late father Donald Wallace's work on a family of sasquatch. For the fifteen years before he died in 2010 he had lived deep in the forest a few miles from the small town of Marblemount, on the banks of the Skagit River. Lori had donated a clump of sasquatch hair found by her father to my research project, and I was keen to interview her and to see the area where the hair had been found.

We left the small town of Marblemount, Washington State,

crossed the bridge over the Skagit River and drove along a narrow road through steep, forested slopes, only now and then glimpsing snow-covered peaks through gaps in the trees. After twenty miles or so we reached a point where a track led off to the right towards a campsite. We parked the car. It was completely silent. No breath of wind, no birdsong. A locked metal gate closed off access to the campground for the winter. We had to continue by foot.

On the way up to Marblemount, Lori and I had talked about precautions in case of a bear encounter. Black bears were common in the area and, in recent years, grizzlies had begun to drift down from British Columbia across the Canadian border, only sixty miles north. This year, with a mild winter, bears were coming out of hibernation earlier than usual. Opinions vary about what to do when meeting a bear, but Rhett, our other companion, was clearly taking the ultimate precaution as he strapped on his sidearm. All I carried was a puny Swiss Army knife.

We had stopped the car near a patch of old snow (which I checked for prints), eased ourselves around the gate and begun walking down the sloping track towards the campground. The underbrush was a mossy carpet punctuated by clumps of narrow-leaved ferns that had been flattened by recent snow. Tall fir trees stretched a hundred feet or more towards the sky. Beneath, spindly saplings struggled upward towards the light, their branches sleeved in the same green velvet moss that covered the ground. The forest was not dense, and the scene was bathed in an entrancing golden glow. To our right, about fifty yards distant, a small river tumbled down the mountainside and filled the wood with the gentle sounds of rushing water. A fallen trunk lay across our path, axe cuts showing that the park rangers had begun to clear the casualties of winter storms. Both Rhett and Lori examined the trees for signs of sasquatch, pointing out how the lower branches of the mossy trees bent downward, which they both attributed to long-term climbing by our mysterious friends. Similar explanations were given for the angle of other fallen trees and branches. Throughout, I said nothing, and saw nothing about the trees that could not be explained by completely ordinary events. I was just an observer, scanning the forest for signs of life, particularly bears, and keeping an open mind. I felt pleased to discover that although I was certainly alert, I was not unduly frightened. I made sure my voice recorder was working and my camera ready for instant action.

About a mile into the forest we came to our destination, a huge fir tree nearly thirty feet round at its base and well over a hundred feet tall. This, I was told, was the Big Guy's tree and he lived in a cave beneath it. The thought that I was in the company of the insane or deluded did flash across my mind. Lori had seemed perfectly normal when I met her in Burlington, and the three of us had chatted easily enough on the drive up the Skagit Valley to Marblemount. And yet here we were, in the middle of the forest, miles from anywhere, about to disturb what, if Lori and Rhett were right, was a very large and potentially very dangerous animal. Lori told me how she had been building a relationship with the Big Guy over several years, visiting this spot regularly, leaving green apples as gifts and engaging him in two-way conversations. She walked over the mossy ground to the base of the tree, all the time talking to the subterranean sasquatch as if it were a small child, pleading with it to respond. Not getting an answer from her voice alone, she stamped hard on the ground, but nothing happened. I watched this performance, not with silent mocking or wry amusement, but as an open-minded observer, a state of mind I tried to retain throughout the project.

Lori continued her monologue for perhaps five minutes, explaining to the Big Guy that she had brought two friends with her and that I had travelled all the way from England to meet him. In a mildly scolding tone she said how disappointed she was that he didn't want to play the knocking game today. We withdrew perhaps twenty yards up the road, Lori reasoning that a break might put him in a better mood and give him more time to wake up. She began to tell me how she had been introduced to the Big Guy by her father and how they had grown close over the years of her visits. Although she and the Big Guy had never seen each other, she had become increasingly aware of his amorous intentions towards her. She became convinced of this when she brought her new fiancé to the gifting spot, whereupon the Big Guy had responded with more agitated knocking and fiercer growling than ever before.

Two or three minutes later we returned and Lori began her routine once again, stamping her foot on the ground. Again there was no response. Lori had with her a tape recorder with a para-bolic microphone to record the Big Guy's various sounds but which could also transmit. She switched the recorder to 'PLAY' mode and the voice of her father, long dead, drifted through the forest. He was reminiscing about his years in the woods, how he had first encountered the family of sasquatch who shared this remote place, how he had won their trust. Her father's familiar and reassuring voice had worked before, encouraging the Big Guy to respond with a knock or a growl. But not this time.

Then, a few seconds later, I heard, we all heard, two distinct knocks coming from under the tree. And a third.

'Did you hear that?' Lori turned and asked me.

'Yes, I did,' I replied. The sound was quite different from Lori's stamping, which was muffled by the mossy ground. The knocks were much sharper, as if a piece of wood was being struck by something hard. I could also feel a very slight vibration in the air at the same time, which ruled out the tape recorder as the source of the sound. I was completely stunned. My first thought: 'Perhaps there is something after all. Perhaps they were right all along.' My second thought: 'What on earth is making the noise?' My third: 'What would Sherlock Holmes have made of it?' In even the most unlikely and mysterious of events, the master detective was always able to provide a rational explanation. If there was one here, I certainly could not think of it.

Gingerly I circled the trunk looking for an entrance to an underground cavern. There was none. The tree stood on a small bluff, with a drop of about ten feet on the downhill side, but I could not see any signs of trampling in the undergrowth. Around the tree, the ground had been flattened and there was a fallen log stripped of bark, as if by rubbing. A bear could have done this. I began to examine the trunk and the undergrowth for hairs, even the apple store, but could find nothing.

I stood up next to Lori and it happened again. This time I was certain she had not done anything. There *must* be something under the tree. I went around the trunk again searching for hair, or hidden openings. It did occur to me that I might tumble through a concealed trapdoor into the creature's den. Even so, I was not unduly frightened, as curiosity and the prospect of a definitive sasquatch identification reinforced my adrenalin-fuelled bravado. I found nothing. Yet the knocking sound had been absolutely definite. None of us were keen to hang around and, a few minutes later, with no further sounds, we walked back to the vehicle and drove away. Other than two black-tailed deer by the roadside, we saw no signs of animal life.

What was I to make of it? First, there was no doubt at all that I heard a total of six knocks coming from under the tree. Since there were two other witnesses, this was no hallucination. Significantly, I thought, Lori was pleased, though not ecstatic, that I had heard the knocks because she, of course, expected to hear them. Rhett too, though he was coughing badly and too sick to register much of a reaction, was not especially excited.

On the drive back to Marblemount, I began to imagine that I would soon have solid evidence of the sasquatch's existence within my grasp. Writing now, a few hours later, I am not so sure. I had certainly heard something extremely strange. But that did not mean what I had heard was a sasquatch. There might be other explanations. Of these a hibernating bear was the most obvious, though this would be hard to reconcile with Lori's claims that the animal responded to her stamping, nor with the far louder and more boisterous knocking accompanied by agitated, blood-curdling screams which she told me she had heard on other occasions. For now I had to be content that this was a true mystery – something that had no rational explanation – which was, for me, intolerably frustrating. I knew already I would be back.

2

The Yeti Enigma

For two hours we watched them. They were enormous and they walked on their hind legs. Their faces I could not see in detail, but the heads were squarish and their ears must lie close to the skull because there was no projection from the silhouette against the snow. The shoulders sloped sharply down to a powerful chest and long arms, the wrists of which reached the knees. The nearest I can get to deciding their colour is a rusty camel. They were covered with a long loose straight hair. They were doing nothing but moving around slowly together and occasionally just standing and looking about them, like people admiring the view.

This graphic description of a close encounter with a pair of yetis in western Nepal comes from the journal of Slavomir Rawicz, a Polish army officer who escaped from a Siberian prisoner-of-war camp in 1941. He and six companions trekked over four thousand miles across tundra and desert before crossing the Himalayas, where they encountered the yetis, before finally reaching safety in India.[1]

Like many of us, I am thrilled by tales like this from faraway lands. Tales of creatures, half-man, half-beast, that roam the high peaks or survive in the densest jungles. I wasn't sure I believed them, but neither was I ready completely to dismiss them. There *could* be something 'out there'.

I have spent my professional life as a scientist, most of it in Oxford, where I specialised in using DNA to explore various aspects of the human past. In particular I have used DNA to work out how our ancestors spread across the planet, when and where they came from and what routes they took. As well as publishing my research in conventional scientific journals, I have written four books which cover the main areas for general readers. *The Seven Daughters of Eve*, published in 2001, concentrates on tracing our ancestry using the maternally inherited mitochondrial DNA, which also features heavily in *The Nature of the Beast*. Other books focus on the paternally inherited Y-chromosome and the evolution of sex (*Adam's Curse*, 2003), on genealogy and the genetic history of Britain and Ireland (*Blood of the Isles*, 2006) and America (*DNA USA*, 2012). I mention these titles in case readers want fuller details of some of the technical aspects that we are going to cover here, though let me reassure you that it is certainly not necessary to have read any of them to follow *The Nature of the Beast*.

I have always been curious about other human species, like the Neanderthals, that we know lived alongside our *Homo sapiens* ancestors. I wondered what happened to them. Did they become extinct, as most authorities believe, or do they live on as creatures such as Rawicz describes? Until very recently this was an absurd notion, but scientific developments over the last few years, which I shall describe, have come some way to making this less of a whimsical fantasy and more of a realistic possibility.

As I began to think seriously about making a scientific investigation in this area, I was frustrated by how little of any value had been published. I read the regular reports in the newspapers

about mysterious remains being sent away to un-named laboratories for DNA testing but these were hardly ever followed up,
and certainly never published in scientific journals in such a way
that I could scrutinise the results.

As I read more, I also discovered a worrying undertone. In
almost every book written by cryptozoologists, as those who study
creatures 'unknown to science' are called, I encountered the
complaint that they had been 'rejected by science'. As a scientist,
I knew very well that science does not *reject* anything out of hand.
Science is a way of trying to make sense of the world that relies
on evidence. As such science is, at heart, a branch of philosophy,
which is the reason practitioners qualify as PhDs – Doctors of
Philosophy. Science is a philosophy based not on opinion or subjec-
tive judgement or orders from a higher authority or from God,
but on evidence. I felt as though my profession was being unfairly
accused by the community of cryptozoologists.

For a mixture of these reasons, I set out to explore what I call
the yeti enigma using the standard approach of my profession.
I would gather genetic evidence for the existence of 'anomalous
primates', as yetis, Bigfoot and others are collectively known,
have a close look at it and, importantly, try to publish what I
found in a mainstream scientific journal. I was strongly of the
opinion that, bizarre though such a project might appear to be,
it did not lie outside the scope of scientific enquiry.

There are many good reasons for doubting the claims of the
yeti-hunters. No body has ever been found and fully examined.
There are no completely convincing films or photographs of these
creatures, even nowadays when superb footage of extremely rare
animals is on our television screens at regular intervals and
everyone has a mobile with a built-in camera. And yet eyewitness reports of these creatures still come streaming in. Are
these all the invention of vivid imaginations, phantasms of
the mind of the harmlessly deluded or just plain fraud? In August
2012, forty-four-year-old Randy Lee Tanley, dressed in a monkey
suit, was run over and killed on Highway 93 near Kalispell,
Montana when he jumped out in front of a car. How many times

had his dangerous antics triggered a new report of a sasquatch sighting from a bewildered and frightened motorist?

What would it take to convince us all of the almost miraculous existence of these creatures? In Scotland, Edinburgh's Royal Mile runs up a gentle slope in the Old Town between the Palace of Holyroodhouse and Edinburgh Castle. About halfway up is the seated bronze statue of the eighteenth-century philosopher David Hume. His big toe protrudes beyond the stone plinth and is polished by the touch of tourists flowing constantly up and down the hill. I doubt many of them know much about David Hume, apart from his irresistibly tangible hallux. Hume agonised over the existence of God and wrote an influential essay 'On Miracles' which sets out what it would take for him to believe in one. After insisting on multiple eyewitness accounts and other criteria, he summarises the level of proof required to convince him and, by implication, all those with a rational mind:

> No testimony is sufficient to establish a miracle, unless the testimony itself be of such kind that its falsehood would be more miraculous than the fact which it endeavours to establish.

In other words, the proof would need to be so convincing that for it *not* to prove the miracle would itself be miraculous. That seemed like a good standard to aim for in my examination of the yeti and Bigfoot evidence. If I had doubts, then I only had to imagine myself presenting each piece of evidence to David Hume for his opinion on its value.

Hume also clearly recognised in his essay that rationality and human nature do not always agree when he wrote:

> With what greediness are the miraculous accounts of travellers received, their descriptions of sea and land monsters, their relations of wonderful adventures, strange men and uncouth manners? . . . The *avidum genus auricularum*, the gazing populace, receives greedily without examination whatever soothes superstition and promotes wonder.

He could have been writing about yetis. I see close similarities between the level of proof he insists upon for miracles, given the fanciful inclinations of human nature ranged against reason, and what most of us would need before we believed in yetis or sasquatch or any other anomalous primate. A live capture, a thoroughly investigated body, possibly even a good-quality, unadulterated film or photograph might be enough. But in their absence is there anything else capable of providing such high levels of proof? It is my belief that DNA, if used properly, does have that capability. It cannot be forged, so far as I know, and with the results independently verified, would, I am fairly certain, satisfy even the great philosopher.

This adventure was not my first excursion into the world of anomalous primates. In 2000, I had received three hair samples in my laboratory from the remote Himalayan kingdom of Bhutan. They were from the *migoi*, the Bhutanese equivalent of the yeti. I had been asked to identify the *migoi* hairs using modern DNA analysis, in much the same way that I had used these techniques for many years to explore the human past.

The *migoi* hairs did not surrender their secrets easily, but eventually two of them were identified as known species of bear. The third remained a mystery. There was DNA, but I could not identify the creature it had come from. The *migoi* project was a sideline, an amusing distraction from the main work of the laboratory. The unused *migoi* samples joined the thousands of others in the freezer and we carried on with our mainstream research into human origins. But I never completely forgot about the *migoi*.

Ten years later, two scientific developments caused the *migoi* to bubble up into my thoughts once again. The first was purely technical. Our main difficulty in getting DNA from the *migoi* hairs had been that there was very little of it in the first place. Only the hair follicle, the root, contained enough DNA for analysis using the lab protocols of the time. Between then and now the protocols have improved a lot, so that these days an intact follicle is no longer necessary, and I found that I could get a very good DNA signal from a single hair with no root attached. This proved to be the technical breakthrough that made this current project feasible.

The second development was more intellectual than technical and arose from the surprising conclusion of a paper published in the journal *Science* in 2010. This article contained details of the DNA sequence from the fossilised remains of another human species, a Neanderthal, widely thought to be extinct. By comparing the Neanderthal DNA sequence with that of modern humans the researchers had concluded that the genomes of Europeans and Asians, but not Africans, contain a small amount of Neanderthal DNA. The explanation offered was that the ancestors of Europeans and Asians had interbred with Neanderthals. This conclusion supplied an intellectual focus for examining the notion, popular among cryptozoologists, that small groups of Neanderthals had somehow managed to survive in remote forests and mountains until recent times, or maybe even to the present day.

While scarcely guaranteeing success, these two developments – the technical ability to identify the species origin of any hair sample from a single shaft, coupled with the strong intellectual case for interbreeding – persuaded me that I now had the tools to do some proper science in what most scientists, for reasons we will explore later, regard as a taboo field. I certainly would not have contemplated getting involved in this work any earlier in my career. Now I am less concerned about what other people think, and have the freedom to explore avenues of research that would have been foolish when I was younger.

Let me be completely clear. I deliberately did *not* set out to find the yeti. Instead I set a goal to locate and analyse as many hair samples as I was able that had been attributed to anomalous primates, in particular to the Himalayan yeti, the Bigfoot/sasquatch of North America (I use the term interchangeably throughout), the Russian *almasty* and the diminutive *orang-pendek* of Sumatra.

In doing so, I found myself entering a strange world of mystery and sensationalism, fraud and obsession and even, at times, the supernatural. I felt safe in doing so only because I was protected by the ruthless rigour of genetic analysis. I was ready to listen to the stories of enthusiasts and eccentrics, liars and lunatics, without having to form an opinion. The only opinion that mattered belonged to the DNA. I certainly met some extraordinary characters along the way, many of whom you will meet later on – people who have spent their lives looking for these creatures and are utterly convinced of their existence. Any doubt is tantamount to heresy and at least one website devoted to Bigfoot has adopted this quotation from the American economist and social theorist Stuart Chase as their mantra.[2]

For those who believe, no proof is necessary.
For those who don't believe, no proof is possible.

The distinction between this and Hume's rationalism could not be more stark.

Cryptozoologists are the unrepentant advocates for one face of the yeti enigma, with plenty of 'evidence' to back their claims. On the other are the all too obvious holes in their argument and the glaring absence of a single piece of evidence that is universally convincing and accepted. It is this enigma I set out to explore.

3

The Last Neanderthal

It could not be described as an extraordinary death. Just an old man dying alone. And yet it was at that moment that we lost our last chance of knowing, really knowing, our nearest human relative. For this man was the last of a dying species, much closer in both genetic and cultural terms than any strained comparison with the chimpanzee, gorilla or orang-utan could ever be. Though he did not know it, this man was the last Neanderthal. His death marked the moment that he, and with him his entire species, became extinct. From that moment on we, *Homo sapiens*, became the only human species on the planet.

The location for this unannounced extinction was a cave high up on a limestone bluff above the Mediterranean in what is now southern Spain, not far from the modern city of Malaga. It happened thirty thousand years ago, ten thousand years before the coldest phase of the last Ice Age. His ancestors had ruled a continent for over a quarter of a million years. From Britain in the West to Iran in the East, they had hunted wild game and brought up their children. They had survived conditions much colder than today, more like Greenland than anything on continental Europe. To cope with the severe conditions Neanderthals had evolved to become compact, hairy and immensely strong. For all but the last forty-five thousand years they had the continent to themselves, the only human species.

But then a new form of human appeared from the Middle East; much lighter-boned, almost lithe by comparison. They had great difficulty in coping with conditions in Europe that were so very different from the plains of East Africa where they had evolved a hundred thousand years or more before. The new arrivals were our ancestors, *Homo sapiens*, and they would eventually drive the Neanderthals to extinction. The process was not deliberate, and for twenty thousand years the two human species lived side by side. But slowly the Neanderthals were confined to less and less productive territories. They had to spend longer and longer hunting to catch less and less. They seemed for some reason unable to adopt the superior weapons technology of their cousins. Weakened by hunger, most died from starvation rather than from any violent confrontation with our ancestors.

And that is what happened to the last Neanderthal. He was a man of about thirty years, born into a small group of five – his parents, an elder brother and two sisters. Apart from other members of the family, he did not see another Neanderthal during his entire life, and neither had his parents, who were actually brother and sister. Weakened by the genetic effects of inbreeding, his own sisters died young and his brother had perished while hunting, a common enough occurrence. His parents passed away when he was in his twenties. He was alone.

Unable to join up with others to form a hunting party, he scraped a living by scavenging the carcasses left behind by predators, including the relative newcomers, our own *Homo sapiens* ancestors. He took trouble to avoid them at a kill, keeping well out of sight behind whatever cover he could find until they had taken what they wanted and returned to camp. He also faced competition from other scavengers. Foxes and vultures were easily coped with, but when he heard the whinnying cry of a hyena, he hastily withdrew. The hyenas, fast, intelligent and with their fearsome bone-crushing jaws, would have killed him without a second thought. He managed to maintain this lonely existence for ten years. He thought of leaving to search for others of his kind, but he also knew that it would be risky to try to survive in unfamiliar territory. What he did not know is that his search would have been entirely futile. There were no other Neanderthals. He was the last.

Every year he grew weaker, not least because hunger drove him to scavenge even long-dead cadavers, after which he was often violently sick. Evolution had not prepared him for this life, unlike the vulture and the hyena who could eat rotting flesh without suffering any ill effects. Eventually he became too weak to leave his rock shelter, a hundred feet up a sea cliff and only accessible by a difficult climb from the shore. Emaciated and unable to move, it was here that he died. Slipping into unconsciousness his eyes closed for the final time on the blue Mediterranean with the afternoon sun shimmering on the ruffled surface of the sea.

I wrote this rather fanciful account of the final demise of the Neanderthals in 2005, when I had been contemplating writing a book on the topic. It comes from the prevailing opinion at the time, that after sharing Europe and parts of western Asia with our *Homo sapiens* ancestors for twenty millennia, Neanderthals became extinct. And when a species becomes extinct, one of them has to be the last survivor. I placed him in southern Spain because it is there, in Zaffaraya Cave not far from Malaga, that the youngest undisputed remains, a mandible, or lower jaw, of a Neanderthal was excavated in 1983. The jaw was dated to 29,550 years BP, the youngest found so far. (*Before Present* is the standard archaeological term for times past, measured against 1 January 1950.)

In the few years since I wrote those words, a lot has changed. Neanderthals had clearly travelled further into Asia than Iran, with remains found twelve hundred miles further east in Okladnikov Cave among the Altai Mountains of southern Siberia in 2007. Yet more human species have been discovered, most notably at

Denisova Cave, also in southern Siberia. What has also changed is what we have found out from genetics. We already knew a great deal from the study of mitochondrial DNA. This small piece of DNA has been a favourite of mine since I and my research team were the first to recover genetic material from ancient human bones in the late 1980s. In ways I will explain in more detail later, mitochondrial DNA has led the way in all explorations of genetic ancestry, from the very recent to the very remote. Its unique pattern of inheritance, being passed down only through the maternal line, means that it traces our matrilineal genealogy, from mother to mother, virtually unchanged from the present day back for thousands of years into the deep past. I have used mitochondrial DNA extensively in my research, both from ancient human fossils and living people, to trace the origins of, among others, the Polynesians, early Europeans, the British and most recently, the Americans.

As far as our genetic relationship to the Neanderthals is concerned, by 2005 we had the genetic fingerprints of mitochondrial DNA recovered from a handful of Neanderthal fossils. These showed that although Neanderthals were certainly related to *Homo sapiens*, they had not been our direct ancestors, which many had once thought. Five years later, the arduous business of sequencing the Neanderthal nuclear genome had been completed. To everyone's surprise this work came to the conclusion that, in Europe and Asia but not Africa, a small but significant proportion of our DNA had been directly inherited from Neanderthal ancestors through interbreeding.

In a small way, the Neanderthals live on in many of us, but I still wondered about their practical extinction. Largely insulated from our own mortality, it is hard for most of us to imagine that whole species can disappear. But, of course, they can and do and have done since the dawn of time. So the extinction of a human species is not in the least remarkable, yet it has necessarily to be accompanied by a final death of the sort I imagined happening in Zaffaraya Cave. Somewhere, at some point, there had to be a last Neanderthal.

I never did write that book. I wondered, only very vaguely,

whether I could ever find a modern human with a Neanderthal mitochondrial DNA, which would offer instant proof that they had bred with our ancestors. The closest I got was a very unusual skull that I came across, entirely by chance, in the museum of the Royal College of Physicians in Edinburgh. In 2008, my son Richard was studying for his A-Levels at the Edinburgh Academy and was doing an art project involving anatomical drawings. I visited the museum to look out some suitable material for his project. During this visit, I got talking to the curator Andrew Connell, who told me about the 'Neanderthaloid' skull.

It had formed part of the collection of the late Dr David Greig. He had an interest in abnormal skulls, many of which were displayed in the museum. The 'Neanderthaloid' skull was one of these. Greig went as far as writing a monograph on it which the College published in 1933.[1] Andrew Connell got a copy down from the shelf. It contained some masterly drawings of the skull, whose accuracy was confirmed when I examined the skull itself. However, what the monograph does not contain are any details of the skull's provenance or how it came to be in Dr Greig's collection. Andrew Connell thought this highly unusual for the otherwise meticulous Greig and suspected this lack of detail masked an unsavoury provenance of some kind. Edinburgh was after all the home of the notorious grave-robbers turned murderers Burke and Hare and of their patron Dr Robert Knox, who received the bodies for dissection. It is entirely my speculation that the 'Neanderthaloid Skull' might have been added to the collection in this way, but the silence about its origins was certainly extremely unusual for Greig.

When I had a close look at the skull the brow ridges were remarkably prominent – hence its labelling as 'Neanderthaloid'. But other parts of the face were less like a Neanderthal. For one, the cheeks were not pushed forward and there was a fairly prominent chin. There were numerous bony lesions on the surface indicative of pathology, possibly tuberculosis. Greig's monograph, which contains a very detailed description of the skull, does not come to any conclusion on the significance of the brow ridges or the surface irregularities. The skull had been cut in half to reveal the inner cranium and the section revealed a thick rim of parietal bone at the level of the cut. There were four teeth, none of them in good condition.

The first thing I wanted to know was the age of the skull. Had it been over thirty thousand years old it probably was a genuine Neanderthal, interesting but not unique. But if it were much younger then it would have been more interesting, a surviving Neanderthal or a hybrid perhaps. It looked quite modern, but to be sure I needed to carbon-date a sample.

After considering my application, the College kindly allowed me to drill into the parietal bone and I sent this off to my colleagues in Oxford for dating. It certainly did turn out to be modern, with a radiocarbon date of only 270 years. To see if it was from a surviving Neanderthal I drilled out another section and sent that to Adelaide for DNA extraction and mitochondrial DNA sequencing. The extraction was successful and the sequence was a good one. It was not Neanderthal however, but sub-Saharan African.

I discussed the results from the Neanderthaloid skull with Prof. Chris Stringer from the Natural History Museum in London. Chris is the foremost expert on Neanderthal fossils and someone with whom I have worked on other projects in the past. We talked about the feasibility of finding Neanderthal DNA that had survived into modern times and wondered where best to search for it. Finding Neanderthal mitochondrial DNA in just one living *Homo sapiens* would be sure proof of interbreeding between the two human species. Chris suggested three locations: southern Spain around Zaffaraya but also the Caucasus region and the Altai Mountains of Siberia. The youngest Neanderthal was found in Spain and the other regions are known hotspots for Neanderthal fossils and ones where Chris thought I would also have the best chance of finding a living person carrying the mitochondrial DNA of a Neanderthal. I carried on with other work, though still keeping an eye open for people with the characteristic features of the Neanderthal: the flat forehead, prominent eyebrow ridges, receding chin and so forth. I saw plenty of candidates, but none so remarkable as in Paris, at the Musée de l'Homme.

I had been in Paris for a few days on business and decided to call in to the museum, a short way across the Seine from the Eiffel Tower. There was no particular reason for the visit other than I had not been there for a long time. The Musée de l'Homme is the Parisian equivalent of the Natural History Museum in London, but with a focus on human evolution, as its name suggests. On the first floor of the museum is a diorama of life-size models of our earliest 'ancestors' starting from the distinctly ape-like Australopithecines, through more advanced forms to modern *Homo sapiens*. The models complement the museum's fabulous collection of human fossils displayed in glass cabinets nearby. It was a Saturday afternoon, and the museum was full of children scampering to and fro, occasionally looking at the displays though more often just running the length of the hall and back again. As I worked my way along the diorama through millions of years of evolution, I eventually came to the penultimate model in the series, the sturdy figure of a Neanderthal.

Standing in front of the display was a man. He was staring at the model with an almost tangible beam of intense attention. He stood completely still and I wondered at first what he had seen in the display that I had not. Something told me he did not want anyone to disturb his concentration. I withdrew to display cabinets a short distance away, from where I watched this strange visitor. He remained where he was for several minutes, without moving, utterly transfixed by the model in front of him. I could only see his back, but he was about 5'6" tall, with wavy, dark hair just covering the collar of his raincoat. After some time he turned to his left, and for a moment I could see him in profile. Even though the light was dim, there was no mistaking the forehead sloping back at about forty-five degrees from the rest of his face. And below that, the strong, protruding brow ridges and a receding chin. I could not escape the chilling sensation that this man, whoever he was, was a Neanderthal who was communicating with the spirit of his ancestor. When I looked again, he was gone.

*

There have been other occasions when I have been alerted to the continued existence of Neanderthals. Even as far back as 1996 when my colleagues and I published the first extensive mitochondrial DNA studies on Europe, one of the minor conclusions was that, on the evidence we had gathered, there was no place for Neanderthals in the ancestry of modern Europeans. The less cerebral British newspapers thought this was clearly wrong. The *Daily Express*, for example, announced in an article entitled 'I'm a Neanderthal man' that 'despite what the scientists say, it explains why men have been behaving badly for the past 100,000 years.' Beneath the headline, the article illustrated the evidence for the persistence of Neanderthals right down to the present day, with portraits of the cricketer Ian 'Beefy' Botham and the former Oasis singer Liam Gallagher looking particularly gormless. Beside them was one of 'Evolution's Survivors' as they put it, the bodybuilder, Hollywood star and former Governor of California, Arnold Schwarzenegger. These were given as examples of what the Neanderthals have become in the popular imagination. Brutish, primitive and rather stupid.

The *Daily Express* article, and others like it that appeared around the world, sparked a deluge of letters from members of the public who refused to believe that the Neanderthals were no longer with us, usually because they knew one. I particularly remember one letter from Santa Barbara, California, which informed me that the checkout clerk at the Safeway store on State Street was definitely a Neanderthal. He, my correspondent assured me, was very nice and would not mind in the least if I asked him for a DNA sample. One revealing aspect of this flood of intelligence was that not once did anyone write to me saying that *they* were a surviving Neanderthal. It was always someone else.

After exploring questions about European ancestry on a timescale of tens of thousands of years, I turned my attention to more recent events, in particular the genetic history of Britain and Ireland, which had been my interest behind an earlier book *Blood of the Isles*. I only mentioned Neanderthals once in that book, and even then only as a narrative sideline rather than as part of my genetic conclusions. Yet when the book was published in 2006 this almost casual remark led to another flood of correspondence. To gather the genetic evidence for *Blood of the Isles*, my research team and I had travelled to every corner of Britain collecting DNA samples from over ten thousand volunteers. In *Blood of the Isles*, I had recounted a story told to me in Mid Wales by a local farmer in the market town of Tregaron. He had lived thereabouts all his life and when, during conversation in the bar of the local hotel, I explained what I was doing in the area, he told me about the Tregaron Neanderthals. Apparently, two elderly bachelors living on a remote farm in the wild hills behind the town were widely acknowledged to be Neanderthals, so much so that the local school organised annual trips to visit them. I wasn't sure I believed him, but nevertheless included the encounter in the book. When the widely read *Wales on Sunday* reviewed *Blood of the Isles*, the article was almost entirely devoted to the Tregaron Neanderthals. There were several consequences of this narrowly focused coverage including, to my bewilderment, that the story had mutated almost immediately into my actual discovery of genetic evidence of Neanderthal survivors in Wales, which of course I had not.

The following week I received two letters from people who had read the *Wales on Sunday* article. Each described two brothers with very unusual heads whom they remember visiting as children in the 1950s. These brothers lived not in the hills behind Tregaron, but a few miles further north under the shadow of Plynlimon, the mountain inland from Aberystwyth that is the source of both the Wye and Severn rivers. Their home had since been flooded to create the Nant-y-Moch reservoir. According to my correspondents, one of the brothers had left his body to Aberystwyth University. These were old memories, and while they may not be entirely accurate, they were nevertheless sufficiently vivid to be recalled fifty years later. In my research for *Blood of the Isles* I had been reading H.J. Fleure's book, *The Natural History of Man in Britain*, published in 1951.[2] Fleure was a distinguished academic, a fellow of the Royal Society and Professor of Anthropology at Aberystwyth. He had made a study of head shapes in Wales and had come to the conclusion that some of the people around Plynlimon might be an archaic form of human. He did not directly suggest that they were Neanderthals, but nonetheless I found it an interesting observation. He illustrated his ideas with a full-face and profile photographs of J. James whose features were typical of Fleure's 'archaic form'. I came to the conclusion that James was one of the brothers who was forced to move when the dam was built and the valley flooded, as mentioned by my Welsh correspondents.

My interest in Fleure's hypothesis and in the Tregaron Neanderthals was reawakened when I was thinking about where best to look for the DNA evidence of Neanderthal survival or interbreeding and I decided, on the spur of the moment, to go back to Mid Wales and see what else I could find out. As I crossed the border into Wales and got closer to my destination, the hill-tops began oddly to assume the shape of a Neanderthal skull, an early indication of my own susceptibility to the psychological phenomenon of pareidolia, something we explore later in the context of sasquatch sightings. I booked into the same hotel, the Talbot in Tregaron, where I had first heard about the Neanderthals ten years earlier. This time my visit coincided with the Rugby World Cup and, as I settled into the bar, the France v. New Zealand quarter-final match was showing live from Cardiff on the large television suspended in the corner. This fixes the date as 6 October 2007. It was a tight game, which France won 20–18.

Wales, and New Zealand for that matter, are passionate about their rugby and the bar was filled with appreciative banter. I was sharing a table with some local farmers, young men who were keenly following the match. During the half-time interval I got talking with them. I told them I was trying to locate the Neanderthal brothers. This didn't produce the looks of astonish-ment that I would have expected in most company. Two of the young farmers had been to school in Lampeter, ten miles south of Tregaron, and had actually been on one of the school trips to visit the brothers. More than that, they pointed to an elderly gentleman sitting in a booth on the other side of the bar. 'There's one of them, over there,' they said.

After a stiffening gulp of beer, I went over to where he was sitting with his pint. He was an elderly gentleman with thinning white hair, quietly minding his own business, and gave me a gentle smile as I introduced myself. It was far too noisy in the bar for explanation, so I arranged to meet him and his brother the next day at their remote farm.

The following day was clear and bright. As I drove out into the hills I wondered what sort of reception I could expect on my unorthodox visit. I arrived at the farm gate and walked up the track to the house, high up on a hillside with a wonderful view across the valley of the River Teifi. Dafydd Jones and his brother John were lifelong bachelors and had taken over the farm from their parents. The farmhouse was spartan but scrupulously clean and they could not have given me a warmer welcome. In no time there was a cup of tea and a slice of home-made ginger cake on the table. I never like to ask someone if they think they might be a Neanderthal, such is their negative reputation. With the Jones brothers I explained that my big survey of Wales had shown up what looked to be very ancient lines of human ancestry in central Wales, which was true, and that I was following this up, guided by Prof. Fleure's earlier observations. Despite their billing in the local school curriculum, the brothers had none of the features that I associated with Neanderthals. Nonetheless I wanted to check their mitochondrial DNA and they were happy to give a sample. They also knew about the James family who once lived in a now-flooded valley, and showed me a copy of a book that gave more details of the family. It was called *Good Men and True: The Lives and Tales of the Shepherds in Mid-Wales* by Erwyd Howells.[3] When I returned to Oxford and read the Bodleian Library's copy I discovered that Fleure's J. James was John James Nant-y-Moch, who, with his brother Jim, had been forced to leave their remote farm-house with its adjacent chapel when the valley was flooded in 1961 to feed the Cwm Rheidol power station. John James moved to Capel Dewi where he died in 1966, aged eighty-five. Howells' book mentions that John's unusual skull was bought by a local museum, though I could not find any confirmation of this. But I had at least identified the owner of Fleure's 'archaic' skull and confirmed the story I had heard ten years earlier about the school visits to the Tregaron Neanderthals. I was hardly surprised when the lab report came back that the brothers' mitochondrial DNA was definitely not Neanderthal. They were descendants of Ursula, the oldest of the seven maternal clans that I had identified in Europeans. An ancient clan to be sure, but still far too young to be Neanderthal.

The following year, I heard a piece on the BBC Radio's *Today* programme which suddenly revived my interest in the possibility of finding surviving Neanderthals. The interviewee was Jonathan Downes. He was due to give a lecture in London later the same day about an expedition that was shortly setting off for the Caucasus Mountains in search of the elusive *almasty*, the local yeti which had in the past been linked to Neanderthal survivors. I couldn't get to the lecture, but my wife Ulla was in London and she did go to hear Jonathan and arranged for us to visit him in Devon, which we did very soon afterwards.

Jonathan runs the 'Centre for Fortean Zoology' or CFZ for short, an organisation named after the early American crypto-zoologist Charles Fort, from his home in north Devon. This was, in fact, the first time I had heard the word 'cryptozoologist', let alone met one. Jonathan is one of those people for whom the phrase 'larger than life' was invented. At once tall and heavily built, with a tangled mane of dark hair and a beard to match, Jonathan is the son of an officer in the British Colonial Service. After spending his early childhood in Nigeria and Hong Kong he was sent to, then expelled from, an exclusive public school in Devon. In an eventful and varied adult life he has had spells as a psychiatric nurse, as a member of the cult art-rock band Amphibians from Outer Space and as the editor of a magazine about tropical fish. Now that he is concentrating on cryptozoology, his output is prodigious. Every month he edits a newsletter and he is also the author of several monographs on the subject.

We arrived at Jonathan's home, an ancient cottage complete with gnarled oak beams, one summer's day in 2008. He and his wife Corinna invited us to lunch and we were joined by other guests, including the zoologist Richard Freeman, who was to lead the forthcoming expedition to the Caucasus. I gave Richard some DNA brushes to take with him to swab the local population, whose mitochondrial DNA just might be Neanderthal, as well as encouraging him to be on the lookout for any *almasty* hairs.

It is a strange phenomenon, but when I have found myself in the company of cryptozoologists, their sincerity and absolute belief in the existence of their quarry begins to rub off. Soon I was also thinking it was only a matter of luck whether or not they ran into an *almasty* on the expedition, not a question of whether or not the *almasty* existed. That was taken for granted. (I have to say that I was less convinced that bad luck alone had frustrated earlier CFZ expeditions in their quests for, among others, the Mongolian Death Worm or the Giant Anaconda of Surinam. 'We went in the dry season,' was Richard's rationalisation of their failure on this last occasion.) As it turned out, Richard's expedition to the Caucasus did not return with any DNA swabs – apparently the locals were Muslims and he didn't think they would approve of giving DNA. And the only hair samples that I was sent looked awfully like pine needles! Once again there was really nothing to build on, so I returned to my research project on the genetic history of America.

After that had been finished off, and the book *DNA USA* finally written, I began to think again of Neanderthals and wondered whether it would be at all feasible to combine my rather whimsical interest in these and other 'extinct' humans with a hard look at the evidence for *almastys*, yetis and the like. The rationale for combining the two approaches was that cryptozoologists live in hope that some of these creatures, as yet 'unknown to science', are actually surviving Neanderthals, or some other archaic hominid, that have yet to succumb to the unstoppable advance of *Homo sapiens*. Perhaps we, the scientists, had all been wrong about the extinction of the Neanderthals and somewhere in the wilderness they lived on.

I rang Jonathan to arrange another visit. The timing was fortunate because the following week was the annual meeting of the CFZ, the intriguingly entitled 'Weird Weekend', where enthusiasts get together for two days of fun and games, and some talks. A speaker had cancelled at short notice, and Jonathan immediately invited me to give a replacement presentation, an invitation I accepted at once. What I was really trying to find out by going to 'Weird Weekend' was whether cryptozoologists were serious about finding proper evidence for the existence of their quarry, or were merely colluding in the continuation of shared delusions long after the rest of the world had made up its mind that it was all nonsense. I have to say that 'Weird Weekend' was a bit of a mixture. When a retired police officer showed huge numbers of slides of UFOs taken on Salisbury Plain and concluded his presentation with one of a mirror that had jumped clean off his kitchen wall and lay smashed on the floor, I began to wonder if I was wasting my time.

I gave my talk straight after the UFO marathon. I had prepared something which covered the DNA tests I had done ten years previously with the Bhutanese *migoi*, then went on to challenge the audience to try their best to gather hairs and other materials that could be genetically tested. Materials, in other words, that could – in theory anyway – provide the extraordinary proofs that were needed to convince a sceptical world. I berated them, gently I hope, for complaining that they were constantly being 'rejected by science'. I don't think I have ever given a talk where the audience has paid more attention. No one spoke or even moved for the forty minutes I was speaking.

At the closing ceremony, Jonathan, in his best ringmaster manner, announced that I had won the coveted 'Golden Baboon Award' – for what I am still not sure. The whole weekend had been lots of fun, but on the long drive back to Oxford my thoughts began to crystallise. There was, I was sure, a willingness, even an eagerness among cryptozoologists to look for proper scientific evidence, but no one had much of a clue how to go about it. Here I could help, because I knew what standard of research results would amount to acceptable proof of their hypothesis that these creatures existed; that is my job as an academic scientist. I also realised that nobody had any idea how to interpret the few DNA tests that had been done on their material so far.

I was very struck by a talk about the *orang-pendek*. This elusive ape-like creature is thought to live in the dense mountain forests of Sumatra. The English translation of its Indonesian name is 'short person'; it is relatively small compared to other anomalous primates, maybe three to four feet tall. According to eyewitness descriptions the *orang-pendek* has a round, ape-like face and is covered in greyish hair. In contrast to the well-known orang-utan or 'forest person' which is predominantly arboreal, the *orang-pendek* walks easily on two legs.

As his talk progressed the speaker, Adam Davies, whom I later got to know well, informed the audience that some mitochondrial DNA tests had been run on hair samples found close to a foot-print track. These tests, according to Adam, showed the DNA to be halfway between human and chimpanzee. When I pressed him in question time, there was no detail, no effort to explain what was meant by 'halfway between human and chimpanzee'. In a regular scientific presentation of the sort I am used to hearing, and giving, that response would be unforgivable. The audience would have torn the speaker to shreds. But, very understandably, neither he nor the members of the audience knew enough about genetics to interpret or even question the laboratory results. I later discovered that the details of the analysis had never been fed back from the labs anyway. All he got was an off-the-cuff remark that he had accepted at face value. I could see that I was going to have some serious discussions with the scientist who had done this work to see if I could at least look at the data on which these conclusions were based.

After 'Weird Weekend' I realised that cryptozoologists had no chance of convincing the world of the validity of their claims on their own. Neither did I think that they had been well served by those scientists who had, from time to time, accepted samples, often collected under very difficult circumstances, and who had not even bothered to return proper reports. By the time Ulla and I got back to our Oxford flat that evening, I had decided to do my best to build a valid scientific project out of all this. You must judge from *The Nature of the Beast* whether or not I succeeded.

4

The Footprint that Shook the World

It is doubtful that any footprint has had such an electrifying effect on the public appetite for romance and adventure as the one photographed in 1951 by Eric Shipton. Shipton was the foremost climber of his day and one of the handful of men who had devoted their lives to the conquest of Mount Everest. Before the Second World War, successive expeditions had pushed higher and higher up the flanks of the tallest mountain in the world trying to find a route that would take them to the summit, and safely down afterwards. Shipton himself had been on most of these Himalayan ventures. Nepal was closed to climbers before the war, so all attempts were made from the north, from Tibet. The aftermath of the Chinese annexation of Tibet in 1950 put a stop to further expeditions from that direction.

Fortunately, the Nepalese government began to issue climbing permits for Everest so attention shifted to the relatively unexplored southern routes. It was Eric Shipton who led the 1951 Everest Reconnaissance Expedition that forced a route through the formidable barrier of the Khumbu icefall, in Shipton's own words 'a wild tumble of contorted ice' which blocked the route from base camp to the Lhotse face, the South Col and from there to the summit. He realised the extreme danger of climbing the icefall, not just from the collapsing seracs, but from the constant risk of ice-avalanches falling from the hanging glaciers clinging to the slopes on either side. This danger remains, and such an avalanche killed sixteen Sherpas in April 2014. Nevertheless, Shipton's demonstration that the Khumbu icefall could be crossed settled the favoured route for the expedition that was to put Edmund Hillary and Tenzing Norgay on the summit of Everest two years later.

After they had mapped the Khumbu icefall, the members of the expedition, having achieved their principal objective, split up into two separate parties. Shipton, along with Michael Ward and Sen Tenzing, set off to explore the Menlung La, a 20,000 foot high pass to the north. On the afternoon of 8 November 1951, the party was descending a gently sloping glacier on the other side of the pass. At around 4 p.m. they suddenly came across a series of tracks in the snow. Some of the footprints were only vague impressions but the rest were sharply outlined individual prints in the thin layer of snow that covered the ice. Shipton took four photographs altogether, two of the indistinct impressions and two of the much sharper footprints. To give a sense of scale, he included in one the ice axe of his companion Michael Ward and in the other, his climbing boot. The prints were about the same length as Ward's size 8½ boots, that is around 12 or 13 inches long, but twice as broad. The edges were sharply defined in the crystalline snow with a broad big toe separated from the other toes, of which there were three or four. It was hard to tell. Sen Tenzing immediately identified them as yeti tracks.

The three climbers followed the tracks down the glacier and wherever they crossed one of the many narrow crevasses in the surface of the ice, the toes appeared to be dug in to gain purchase on the far side. Eventually the tracks disappeared on the grass-covered glacial moraine. Two days later the group was joined by other climbers who had travelled down the same glacier, but by then the tracks were gone, eroded by the wind and the sun.

It was the publication of Shipton's ice-axe photograph in the London *Times* on 7 December 1951, the day after his written account in the same paper, that set the world buzzing with excitement and the anticipation of mysterious creatures roaming free in the land of snow and ice, a land which many had heard of but few had experienced for themselves. Strange and unexplained footprints had been seen before in the Himalayas but somehow this image, alongside the ice axe that not only provided scale but also an immediate link to a world of adventure, captured the mystery and the romance of the yeti in a single frame. This photograph joined that select group of iconic images – the Vietnamese girl burned by napalm, Neil Armstrong's photograph of Buzz Aldrin stepping onto the surface of the moon and Marilyn Monroe's white dress billowing in the breeze from a subway grate in *The Seven Year Itch* – that seemed to capture everything in a single moment. Shipton's shot of the yeti print gripped the public imagination of a generation and ushered in two decades of frantic activity, all aimed at finding the creature that made it – alive or dead.

The first of the big yeti expeditions was sponsored by the London newspaper, the *Daily Mail*. The man with the imagination, and the connections, to secure the funding and get the expedition off the ground was the journalist Ralph William Burdick Izzard. He had been sent by the *Mail* to cover the ascent of Everest in 1953, but had been kept well away from the action by the people from the *Times* who had bought the exclusive media rights to the expedition. However, during his otherwise frustrating trip, he did have a brief conversation with John Hunt, the leader of the Everest expedition. Hunt, who had been chosen as the expedition's leader ahead of Eric Shipton, had been to the region many times before and had himself seen yeti tracks. In 1937 he was crossing the 19,000 foot Zemu Gap that lies between Everest and Kanchenjunga to the east when he saw a double set of prints, side by side, stretching ahead of him. At first he interpreted these as tracks left by two members of a German expedition that was in the area. However, he later discovered that the German climbers had been elsewhere and was left with the intriguing possibility that these were indeed yeti tracks. Hunt conveyed to Izzard his belief that the yeti was a real animal, and that it could be found.

As soon as he returned to London, Izzard set about planning his own expedition for the following year. Its aim was not to climb Everest, nobody was interested in doing that after 1953, but to find the creature that had made the prints. If man's ambition to get to the top of the world had been achieved in 1953, that other legacy of Himalayan adventures, the yeti, was a glittering prize yet to be claimed.

Izzard was one of the sizeable band of dashing yet urbane heroes who were said to have been models for Ian Fleming's character James Bond – in Izzard's case, he was allegedly a source of inspiration for Fleming's first novel, *Casino Royale*. After the war he spent over thirty years on the staff of the *Mail*, with postings all over the world. Even so, it was a hard task to convince the paper's proprietor, Lord Rothermere, to put up the money for the scale of expedition Izzard had estimated the task required. But he succeeded, and by early January 1954 the expedition had arrived in Calcutta en route for Nepal. Rather like Hunt's the year before, this was a huge operation organised along military lines and employing, at its height, an army of three hundred porters. Military tactics also governed the expedition's strategy. It would divide into three separate parties to form a complicated pincer movement, driving any yetis into an ambush. Sadly, despite spending fifteen weeks in the Himalayas, they returned to London with nothing more than some vague sightings and even vaguer footprints. History does not record Lord Rothermere's reaction but the satirical magazine *Punch* greeted their return with these pithy lines:

NOTHING DEFINITE YETI
There are fascinating footprints in the snows of Kathmandu
On a slightly less than super-human scale:
There are numerous conjectures on the owners of the shoe
And the money it has cost the *Daily Mail*.

The failure of the Izzard expedition to photograph or, better still, to capture a yeti did nothing to dampen the enthusiasm of other wealthy romantics, principal among whom was the Texas oil millionaire Tom Slick. Indeed, the *Daily Mail*'s failure, or rather lack of success, meant the great prize was still there to be claimed.

Thomas Baker Slick Jr was brought up in a family of entrepreneurs with business interests ranging from beef to railroads to oil, all over the Midwest. In 1912, his father Tom Slick Sr discovered oil in Payne County, Oklahoma, which earned him his soubriquet: 'King of the Wildcatters'. He capitalised on this and other discoveries in the developing Cushing oilfield by building the railroads connecting the wells. In 1920, Tom Sr and his partners created a new town in Creek County, Oklahoma as their centre of operations. Soon the Slick Townsite Company, the Slick Gas Company and the Slick National Bank opened their doors for business in, you've guessed it, the town of Slick. At its peak over five thousand people called Slick home and the town's prosperity boomed. But like so many other boom towns, when the wells dried up the people left. Slick, Oklahoma is now abandoned with little remaining but crumbling buildings, weed-choked schoolrooms and broken window panes.

With a family background like this, it isn't surprising that Tom Jr thought on a grand scale. His father, Tom Sr, died in 1930 at forty-six years of age, when Tom Jr was just fourteen. He had made a fortune from his businesses but never lived to enjoy the benefits, something his son was determined not to repeat. The day after his father died, Tom Jr was sent east from his home in Oklahoma City to attend the exclusive Phillips Exeter College in New Hampshire. At school young Tom had shown an interest in science, especially biology, which continued when he reached New Hampshire. Contemporaries remarked that he also devoted himself with gusto to the more practical applications of human biology, namely dating girls.

In 1934, his graduation from college coincided with the announcement of an ambitious expedition to Tibet to capture the legendary giant panda. Although dead specimens had reached the West, until 1936 only nine Westerners had seen a giant panda alive. The 1936 reconnaissance expedition was led by Bill Harkness, one of the Standard Oil dynasty that had sponsored Tom Slick Jr through college. Harkness died in Shanghai on the reconnaissance trip, but his widow Ruth, a New York fashion designer, decided to lead the main expedition herself. In 1936, she and the naturalist Gerald Russell encountered and then captured a giant panda cub in Sichuan province, Western China. Su Lin, as the young animal was named, arrived back in the United States and created an immediate sensation. For several months, she lived in Ruth Harkness's New York apartment but was eventually sold to Brookfield Zoo, Chicago. Sadly Su Lin died of pneumonia two years later. His (or as it turned out during the autopsy, her) stuffed body is now on display at the Field Museum of Natural History in Chicago. If only all expeditions to find legendary creatures had such a happy ending – at least for the explorers.

Slick progressed smoothly from college to Yale. During his time there he heard about the sightings of the Loch Ness Monster and, in 1937, shipped his car and his friends to Scotland for a visit. They stayed for a while near Urquhart Bay, where most sightings are made, but saw nothing. Ironically, in the same month that Slick and his group were on the north shore, the Rev William Graham saw a two-humped grey creature cruising along the opposite shore at Foyers.

Tom Jr did well at Yale, graduating in biology in 1938. He lived back in Oklahoma City for a while before moving to San Antonio, Texas in 1939. It was here that he began to use his wealth to create a succession of scientific research institutes to conduct experiments in agriculture, engineering, medicine and the natural sciences. Unlike the eponymous town, one of the institutes founded by Tom Slick in 1947 has prospered and is now among the largest independent non-profit applied research and development organisations in the United States, posting an income of $570 million in 2011.

While Tom Slick was building research institutes in Texas, and overseeing the family businesses, one of the most important events in the whole history of the quest for legendary creatures occurred on the other side of the Atlantic. An obscure Belgian biologist called Bernard Heuvelmans published a book in his native French. *Sur la Piste des Bêtes Ignorées* did not create much of a sensation when it was published in 1955, but when Richard Garnett's English language translation appeared in 1958 as *On the Track of Unknown Animals* it immediately became a classic. Over fifty years later it is still in print and has sold over two million copies. Heuvelmans himself was a complex character, one I got to know very well when I spent many days in his archive in Lausanne. But that is for a later chapter. There is no doubting the influence that *On the Track* had on everyone with even the slightest interest in 'animals unknown to science'. And this included Tom Slick.

By the time he made up his mind to send his own expedition in search of the yeti, Slick had already visited India on several occasions and had talked to local people about the half-human, half-ape creatures that roamed the high forests and snowfields of Nepal and Tibet. His interests were not confined to travel, but had in part a business motivation. He had developed an interest in skyscraper construction using pre-cast concrete and thought, bizarrely, that if he could discover the secret of the Hindu practice of levitation, he might be able to adapt it to raising the concrete slabs into place on building sites. Between these researches, he began in earnest to make the practical arrangements for the first of his yeti expeditions, which is when he heard about Peter Byrne. Peter, the last survivor of the big yeti hunts, now lives on the Pacific coast of Oregon in the northwest United States. He has lost none of his enthusiasm for finding anomalous primates as I discovered when I visited him at his home during my own excursions in the region.

Originally from Ireland, Byrne had left the Royal Air Force after World War II and become a tea planter near Darjeeling in the foothills of the Himalayas, within sight, on a clear day, of the third highest mountain in the world, Kanchenjunga. It was during this phase of his life, on a trip to Sikkim in 1948, that Peter Byrne had seen his first yeti print close to the Zemu Glacier, the site of John Hunt's discovery of the twin set of tracks in 1937. As for many before and since, this was a transforming experience and he determined, like others who undergo a similar encounter, to do what he could to find one of these creatures. He quit tea-planting and started a career as a professional big-game hunter, helping clients to track and shoot mainly tiger, leopard and rhinoceros in the forests of Nepal. This was in the days when these animals were common and views on hunting were very different from what they are today. Peter still leads expeditions to Nepal, although these days his hunting parties are armed only with cameras.

As for so many would be yeti-hunters, a major problem was securing the funding to mount a credible expedition. Byrne's move to Australia in 1954, where he worked as a journalist, gave him the opportunity to begin raising funds among the business community in Sydney. Unlike the large-scale *Daily Mail* expedition of 1954, Byrne's idea was to send a small team of two or three travelling lightly through yeti country. Even so, Byrne was not finding it easy to raise enough money for even this modest expedition. He did however have sufficient funds to return to Sikkim to look for additional evidence to convince his prospective backers that their money would not be wasted. It was during this foray, in 1956, that Byrne first heard about Tom Slick from none other than Tenzing Norgay, Hillary's companion on the summit of Everest. Tenzing had heard of Slick's plans to send an expedition to the Himalayas in search of the yeti and gave Byrne his address in Texas. Byrne wrote at once and received a rapid and encouraging reply.

There followed months of complicated negotiations with the Nepalese government but eventually the expedition arrived in Nepal in March 1957. It was a slimmed-down operation comprising Tom Slick, Peter Byrne and the superintendent of the Delhi Zoo, N.D. Bachkheti, with a support team of seven Sherpas, and forty porters to carry the kit. The region they chose to explore was the country surrounding the Arun Valley in northeastern Nepal. Their reasons were that they had been informed by the head of the Geographical Survey of Nepal, one Col Rana, that yetis were comparatively plentiful in the region and that they were much bigger than those found further west.

The Arun is the largest of the trans-Himalayan rivers, rising in Tibet and cutting through the main Himalayan chain between the peaks of Makalu and Kanchenjunga. The lower levels of the Arun basin are heavily populated but the steep-sided valleys leading to the high peaks are virtually unexplored, even today. They are thickly wooded with dense growth of rhododendron between chir pines, fir trees and native hardwoods. The tree line is at about fifteen thousand feet and gives way to a barren alpine landscape that, when the expedition arrived, was partly covered in snow. It was here that Tom Slick saw his first yeti prints, thirteen inches long and with the clear impression of five toes. He photographed and cast these in plaster of Paris. On the same day, but at a different location, Peter Byrne found a good set of fresh prints that he was able to follow for several miles through forest, including thick stands of bamboo that the creature, whatever it was, had ploughed straight through. Imagine his excitement in thinking that he might be on the brink of coming face-to-face with a yeti. Sadly it became clear to the pursuing Byrne that the creature was moving much faster than he was and that he was not going to catch up with it. He returned to camp exhilarated but also frustrated. Others in Byrne's party who had been ordered to follow the tracks in the reverse direction in the hope of finding the creature's lair found only droppings and a single black hair snagged on a thorn bush.

On the way out from the mountains, Slick did an interesting experiment in the villages they passed through along the route. Descriptions of native encounters with the yeti are often tinged by the suspicion that they are confusing ordinary animals with the mythical yeti. Top of the list are bears and, in the Himalayas at least, langurs – large monkeys that are certainly capable of walking on two legs for considerable distances. In Kampalung, for example, he found fifteen villagers who, in the past, claimed to have had a clear view of the yeti in good light. He showed these witnesses a series of twenty photographs or drawings of animals often thought to have been confused with the mythical creature. He asked each witness in turn to pick out the image that most closely resembled the animal they had seen. In every case, the first choice was a photograph of a gorilla, unknown in the Himalayas and only found in Africa. Second choice was an artist's drawing of a prehistoric human and third was an orangutan standing up. No one chose the images of bear or langur, which were immediately recognised for what they were.

Slick and his companions left Nepal after having spent five weeks in the country. Although Slick remained enthusiastic about the chances of finding a yeti, and financed two further expeditions, he never himself returned to Nepal. No one seems to know what happened to the hair, the droppings and footprint casts. There are rumours that they are held to this day in the vaults of *Life* magazine, which covered the story, or kept in the private Slick family archive well away from prying eyes, and nosey scientists like me.

Despite Slick's decision not to go on any more yeti-hunts to Nepal himself, he organised and, with his business partner Kirk Johnson Sr, financed two further forays into the Arun Valley. Once again, Peter Byrne was involved, this time with his brother Bryan, another hunter, and Gerald Russell the naturalist, who had been a member of the successful Harkness giant panda expedition of 1936. The expedition arrived in the Arun Valley in February 1958 and stayed for several months. The strategy was to split up into small groups, find a promising location and to stay there. This time the expedition came with dogs, American Bluetick Coonhounds that were bred for hunting and had the reputation of forcing animals up into trees, where they could then be shot. Blueticks were used widely for hunting jaguar, mountain lion and bear in Central and South America, so they seemed to be ideal for doing the same with the yeti in Nepal. Not that Slick wanted a yeti killed, just tranquillised and captured alive. However, it soon became obvious that the Blueticks were not cut out for high-altitude tracking. They became morose and disobedient, needed constant medical attention, kept escaping and were finally 'discontinued' as the expedition log puts it.

The 1958 expedition did find footprints, a nesting cave and some droppings, but had little more to show for months of arduous work. I would have expected hairs in the nest cave but, if they were taken, they have – like everything else from the Slick expeditions – vanished.

The following year, 1959, Slick and Johnson financed the last of the three yeti-hunts in the Himalayas and sent orders to Peter Byrne, by then in Kathmandu, to return to the mountains. Though the earlier expeditions had been deliberately small, the third was positively minimalistic, involving only Peter and his brother Bryan. The pair had no tents and very little equipment, the rationale being that this was the best way to get close to a yeti. They lived off the land, sleeping in the open or in caves when it snowed. Altogether they spent nine months in Nepal, but still they found only footprints. However, the 1959 expedition did do one thing that was to become the focus of future attempts to identify the yeti. At the monastery in Pangboche, not far from the Everest Base Camp, the Byrne brothers were able to examine and photograph a yeti scalp, and a wizened hand belonging to one of the creatures. These were holy relics, and although the *Daily Mail* expedition of 1954 had seen one of the scalps, they had not been allowed to examine or photograph it.

The Slick expeditions came to a sudden end in the winter of 1959. The Byrne brothers were living in a cave high up in the Chhoyang Khola, one of the steep valleys running down to the Arun. There was deep snow cover and the temperature outside had dropped to 45 degrees below zero, which is about the same on both Fahrenheit and Celsius scales. Their equipment all but lost or destroyed and their clothes in tatters, they were subsisting on *champa*, a local grain mash, yak's milk and what edible ferns they could find beneath the snow. As they were sitting at their campfire a runner arrived from the south. In his hand was a letter from Tom Slick with a fresh set of instructions. As we shall see a little later, as one adventure came to an end, another was about to begin for Peter Byrne.

Although 1959 marked the end of Tom Slick's expeditions in search of the yeti, there were further Himalayan expeditions that set out with the same purpose, the most famous of which was led by Sir Edmund Hillary in 1962. Writing now in New Zealand, I am only too aware of the tremendous regard everyone has for one of their few truly international heroes. His craggy profile decorates the five-dollar note and there are numerous exhibits around the country celebrating not only his conquest of Everest in 1953 but also his subsequent expeditions, such as to the South Pole in 1958. Perhaps most praise for Hillary is reserved for his founding of the Himalayan Trust, which has helped to build schools, roads and hospitals for the Sherpas of Nepal. He is remembered for his modesty and lack of pretension even now, but there is no doubt he became an influential international figure after his triumph on Everest. Awarded a KBE within eight days of reaching the summit, and thereby entitled to use the title 'Sir', he continued to be heaped with honours for the rest of his life.

I was privileged to meet his widow Lady June at her home in Auckland in 2013. I especially wanted to ask her about the blue bear skins and other artefacts that her husband had brought back with him from Nepal. She found some photographs of the skins in one of his books, all special leather-bound and autographed editions, and told me what she could about them. My most vivid memory of the visit is the collection of framed photographs on the sideboard, much as any family might have. Except that these showed Sir Edmund and his family in all stages of his life, from the summit of Everest, to Antarctica, to the official residence in Delhi when he was appointed New Zealand High Commissioner in 1985.

I recognised another famous face in a photograph taken with Reinhold Messner, the first to climb Everest solo and without oxygen and also a man with a deep interest in the yeti, as we shall see. The image on Lady Hillary's mantelpiece was a publicity photograph for a luxury brand of wristwatch. She pointed out

that her husband was a very tall man, so to equalise the height
of the two mountaineers the directors of the shoot had dug a
hole in the snow for Sir Edmund to stand in. When you knew
what to look for it was obvious but, if not, the shot looked perfectly
natural. I felt, just as I did after my encounters with Reinhold
Messner and with Peter Byrne, that I had somehow brushed
against a vanished and heroic world.

The only community that does not regard Sir Edmund as a
hero is the tight-knit group of cryptozoologists. His 1962 expedition to Nepal left with the stated purpose, like others before, of
looking for evidence of the yeti. Unlike Tom Slick, Hillary was
not much interested in the creature himself but wanted to go back
to Nepal to conduct experiments on high-altitude physiology and
the effects of altitude on blood oxygen levels. This aim failed to
excite financial backers, so Hillary was persuaded to include some
yeti-hunting in the prospectus. As soon as he did that, the money
rolled in.

In contrast to the later Slick yeti-hunts, Hillary's 1962 expedition was on a large scale, numbering six hundred people at its
height. They found the usual footprints, though Hillary dismissed
these as the impressions of regular animals enlarged by melting.
For instance he describes following one such set of giant footprints in deep snow only to find that when the tracks passed
through the shadow cast by a ridge, the toes and heel marks
resolved into four separate pugmarks of a small quadruped the
size of a fox.[1] However, like Peter Byrne, Hillary did examine a
yeti scalp, this time at Khumjung, and persuaded the monastery
to let him borrow it for six weeks in order to have it properly
examined in the West. Along with its Khumjung guardian,
Khumjo Chumbi, the scalp toured Europe and America where
the experts soon declared it to be a fake, probably fashioned
from the skin of the serow, a Himalayan goat-antelope.

Neither Hillary nor his companions were impressed by another
holy relic – the yeti hand kept in Pangboche monastery, which
they also examined, though did not borrow. Though attributed
to the yeti, Hillary thought the hand was 'essentially human,

strung together with bits of wire with the possible inclusion of several animal bones'. He was at least partly right. What Hillary did not know was that Peter Byrne had surreptitiously replaced one of the fingers with a human bone during a clandestine visit to the same monastery in 1959. Peter told me the colourful details of this daring heist and I shall relate them in a later chapter.

Not only did Hillary fail to find any convincing evidence for the yeti, he went further than Slick ever did by declaring that the yeti did not exist, much to the disgust of cryptozoologists. It was this further step that cost Hillary the otherwise universal admiration with which he is remembered.

5

The Professor

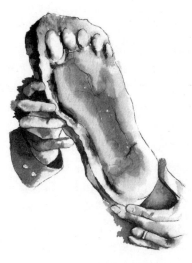

Before I began my own research in earnest, and while I was still finding out was going on in cryptozoology, I was soon referred to Dr Jeff Meldrum from Idaho State University in Pocatello. Dr Meldrum is a professor of anatomy and anthropology and one of the very few full-time academics who are working on anomalous primates as part of their professional activities. As a result, he is in demand all over the world whenever an opinion on Bigfoot or yetis is required. Dr Meldrum's particular expertise lies in the evolution of bipedalism, the art of walking on two feet, and the substantial anatomical changes that our ancestors needed to undergo in order to develop that ability. With our feet securely hidden from view in shoes and boots, we are blissfully unaware of how sophisticated a structure they really are. In his book *Sasquatch*, which I read as soon as I could get hold of a copy, Dr Meldrum explains that in his opinion the foot is second only to the brain in terms of complexity.

By lucky chance my book tour for *DNA USA* took me to Salt

Lake City, about two hours' drive south from Pocatello, and Dr Meldrum very kindly agreed to make the trip down and meet Ulla and me at the Market Street Grill. When he declined a coffee I soon discovered that Jeff was a Mormon, though he didn't mind our own indulgence. I began by asking him how he became professionally involved in sasquatch. He replied that it was triggered by his experience as a young man when he came across a clear set of large prints while out hiking.

In that sense, Dr Meldrum has much in common with many other enthusiasts. This time, however, the experience had a profound influence on his choice of career. After a degree from Brigham Young University in Salt Lake City, a PhD at the State University of New York and postdoctoral work at Duke and Northwestern universities, Dr Meldrum took up an appointment at Idaho State where his interest in sasquatch and other anomalous primates was allowed to flower. Dr Meldrum has persisted in honing his expertise in bipedalism and especially in the analysis of footprints. He has amassed a huge collection of casts from all over the world, and not just of sasquatch and the like.

Unfortunately, there wasn't time on that occasion to ask him much about sasquatch footprints. In any event I was only beginning to find my bearings in this new field and didn't really know what questions to ask. Nevertheless Dr Meldrum agreed to put the word out that I was on the lookout for hair samples that had been attributed to sasquatch and Bigfoot, and later helped me to screen for the most promising specimens. So far we have not met again, so it was left to my researcher Marcus Morris to ask him the more detailed questions that I had formulated. I am including Dr Meldrum's remarks here not because they impact directly on the DNA work, but because footprints are such an important component of the evidence for yeti and Bigfoot. The first thing I wanted to hear about was his opinion of the Shipton 1951 photograph that had such a huge influence in bringing the yeti into the realm of public consciousness.

Dr Meldrum began by highlighting the things that might

happen to a footprint in the snow. The first, of course, is melting, which tends to enlarge a print, or sometimes amalgamate several prints into one large one. As Hillary also pointed out, in certain examples four separate prints of a fox bounding over the snow could be condensed into one large print. Melting could also leave a small pool of water in the print, which can enlarge it at its lowest point, usually the heel, exaggerating these features. This process typically leads to the rounding out of the heel impression which, in the case of a bear for example, transforms it from the typically narrow heel to a broader, more human-like shape.

Ice doesn't have to melt before it disappears. The process of sublimation converts solid ice and snow to water vapour without an intermediate liquid step. You can see sublimation at work if you hang out the washing on a very cold day. The clothes freeze and stiffen, but still dry off even though the temperature never climbs above zero. The sublimation process is speeded up at high altitude, where the pressure and humidity are both low. So it's certainly a factor. Sublimation tends to increase in sunlight, so the shadow cast by the side of a footprint slows sublimation while the opposite side, exposed to the sun, evaporates more quickly. Depending on the angle of the print relative to the sun's rays this can either broaden the impression if the track is at right-angles, or lengthen it if the track is in line with the sun. In snow, the casting process itself may also distort the impression, often breaking down the small ridges of snow that separate the individual toe prints from each other, making it look as if the creature has fewer toes than it really does. So given these difficulties of interpretation, what did he think of the famous Shipton print?

'For a start, it's a very clear footprint in a thin layer of snow in ice. We get the impression of a large, very broad rounded heel and then this odd arrangement of toes. When you consider the entire photograph, though, it is remarkably similar to a melt-out area. You see just a little hint of it where there was some irregularity that was catching the sunlight at its incident angle, and that caused this crescent shape around the heel. The heel is the main support; that's where the pressure is concentrated beneath the foot. And when we look closely at the photograph, to see a ridge running right through what is interpreted as the heel with loose shards of ice or snow scattered there is just inconceivable to me. However, if we say that the deepest point of this area is the heel and we treat the crescent as an artefact of melting and remove it, then what's interesting is that the foot has a better symmetry with a tapered heel which is much more typical of a great ape.'

Dr Meldrum also explained that the same editing to remove melt artefacts enlarges the gap between the big toe and the other digits, which is also typical of a great ape like a gorilla or a chimpanzee. He then turned his attention to the second toe.

'The curious thing that has often sparked a lot of discussion and debate is the second toe. It's been described as something almost like a natural piton, used to help the creature climb up crevices in the rock and so forth. What struck me is that it bears a remarkable resemblance to a condition in human pathology known as macrodactyly. This is a condition where the skeletal elements, but especially the soft tissue, hypertrophy. They enlarge, pathologically. And so you get this over-sized toe and in this case probably some deformity of the first digit as well.

'It's hard to draw very firm conclusions from the Shipton print and, I guess, very unfortunate that under the extreme circumstances they found themselves in, Shipton and Ward only took a photo of a single footprint. But it could be great ape or other hominid. I'm just not sure.'

Did Dr Meldrum think the print was genuine? Sir Edmund Hillary, who was also a member of the 1951 expedition, once quipped that Shipton was a well-known practical joker. But Jeff dismissed the suggestion that the print was a deliberate hoax. For one thing, Shipton was not alone, but accompanied by Michael Ward, whose ice axe appears in the photograph.

'I find that suggestion hard to accept when Michael Ward, a surgeon, a professional, recently published an article in a medical mountaineering journal revisiting the whole issue and discussing possible explanations for the odd morphology of the feet.[1] He didn't necessarily advocate the existence of the yeti but he was trying to explain what could have accounted for these footprints. He even included photographs of Sherpas with deformed feet. One individual had a big toe that stuck out at right angles to his foot. I can't imagine that Ward would perpetuate a practical joke by publishing such a paper. So I think what they saw and photographed, they really did see.

'It's not unreasonable to suggest that this could have been left by a hominid. One of the things that reinforces that inference, that hypothesis, is the presence of another very hominoid-looking footprint that has much more thorough documentation than the Shipton print. This was one that was described from the expedition by the biologists Edward Cronin and Jeffrey McNeely above the Arun Valley in Nepal in 1972.

'They discovered footprints outside their tents in the morning hours before the sun even touched them. And they backtracked. The creature, whatever it was, had come up a very steep slope through deep snow, never touching the ground with its fore limbs. It came marching up there, apparently spied their tents and detoured to meander through the camp before continuing on over the pass it was using to get to the adjacent valley. Cronin and McNeely lost the trail in the rhododendron groves on the other side of the pass.

'Here was a long line of crisp, fresh footprints unaffected by melting or sublimation that was observed and photographed by two professional zoologists. Cronin and McNeely also made a cast of the track but unfortunately it was confiscated at the border as they were leaving and now it's lost. But based on the photographs of the original footprint, it looks remarkably like a chimpanzee, though it cannot be from a chimpanzee because they are confined to African rainforests. No bear has an opposable digit like this creature. Based on that and using reference material from other hominoids, again assuming for a moment that the slight indications of digits and the length of those digits can be interpreted, they can be fitted to a great ape foot. That is what I came up with.

'You know, it's always surprised me that the Cronin-McNeely

footprints have not made a bigger splash amongst the scientific community because here is a biological team up there with the express purpose of studying the wildlife.'

That answered a lot of questions about the Shipton image, and tracks left in snow generally. Other than the Shipton-Ward and Cronin-McNeely prints, Dr Meldrum didn't think any of the other Himalayan tracks could be attributed to hominids. The great majority are so distorted by melting and sublimation that they cannot be identified – but that certainly doesn't mean they are from unidentified animals!

The same uncertainty does not extend to sasquatch/Bigfoot prints, most of which are left on solid ground and are therefore not vulnerable to melt distortion. Dr Meldrum has examined hundreds of footprints from the US and further afield. Just like the Himalayan examples, many are too altered for easy inter-pretation. However many are not, and Dr Meldrum firmly believes that, after filtering out the fakes, there are just too many good prints showing the right anatomical features to come to any other conclusion than that these creatures, whatever they may be, are real.

As he concedes in his book *Sasquatch: Legend meets Science*, footprint evidence alone is never going to be enough to provide the unambiguous proof that these creatures exist. Nonetheless, the widespread occurrence of footprints, while not a sufficient proof in itself, certainly contributes to the creation and persistence of the enigma surrounding the Himalayan yeti and his American cousins.

6

Desperately Seeking Sasquatch

It was Peter Byrne who first cemented the connection between the mysterious Himalayan yeti and its North American counterpart, the sasquatch – or, as it is now much better known, Bigfoot. The letter from Tom Slick that the runner brought to Peter Byrne in his cave-shelter high above the Arun Valley in late 1959 instructed him to terminate the hunt for the yeti in Nepal with immediate effect. The same letter also invited him to consider switching to another elusive quarry. Slick was asking Byrne to drop the yeti in favour of finding Bigfoot. Byrne accepted at once, paid off his Sherpas and, with his brother, trekked back down the mountains to Kathmandu. Within three weeks he was in the small town of Willow Creek, Humboldt County, northern California. A very different country, very different people, though with much the same stories of mysterious primates roaming its montane forests.

Andrew Genzoli of the *Humboldt Times*, the local paper of Willow Creek, was the man who coined the name 'Bigfoot' when, in October 1958, the discovery of giant footprints in the mud at nearby Bluff Creek was first reported in the paper. The story, and the name, soon gained international attention when it was picked up by the Associated Press wire service. The find was made by a 'catskinner' – a bulldozer driver – by the name of Jerry Crew who discovered dozens of prints surrounding his machine when he came to work one morning. Each was about sixteen inches long and seven inches wide with the clear impressions of five toes. He made a plaster cast of one of the prints and, rather like Shipton's 1951 image from the Himalayas, the newspaper picture of a bespectacled Crew holding the giant footprint cast had an immediate impact. In an instant, America had its own mysterious giants 'unknown to science' and they had a name that no one could forget. It hardly matters that years later it emerged that Jerry Crew had probably been hoaxed.

The capture of a Bigfoot on home soil was obviously going to be much less of a major undertaking than sending an expedition to the Himalayas, and a lot less costly, a combination that Tom Slick found irresistible. Hence his letter to Peter Byrne. However, unlike the Himalayas where yeti-hunts were confined to well-financed expeditions, usually with professional naturalists taking part, in America anyone could go looking for Bigfoot – and many did, and still do.

Just as Shipton's was not the first indication of the yeti, neither were Crew's footprints from Bluff Creek the first sign that there might be a large biped in the forests of North America. Just as the yeti and its counterparts in the Himalayas were well known to the indigenous people and were deeply embedded in their cultural mythology, the same was true of the Bigfoot in America. Native Americans from northern California, Oregon, Washington and across the Canadian border into British Columbia all had their own stories of hairy creatures, half-man and half-animal, that lived deep in the forest. They had various names that were compounded by Canadian writer J.W. Burns in the 1920s to *sàsqˋets*, the indigenous Salish term for 'hairy man'. '*Sàsqˋets*' very soon became 'sasquatch'. It was a Native American who provided the first graphic account of his encounter with, and abduction by, a family of sasquatch.

In the autumn of 1928 on Vancouver Island, British Columbia, a Nootka Indian trapper called Muchalat Harry set off on a hunting trip to the mouth of the Conuma River on the west coast of the island. We are indebted for the account of what happened next to Father Anthony Terhaar, a Benedictine monk who knew Muchalat Harry well and wrote down what he was told. Harry left his canoe at the mouth of the river and headed upstream by foot for about twelve miles and made his camp. That night he was woken by being lifted bodily, in his blankets, by a large creature and carried off into the forest for, he guessed, about three or four miles. At daylight he found himself under a rock shelter and surrounded by twenty gigantic hairy creatures standing on two legs. They were not acting aggressively towards Harry, just standing and staring at him. His fear turned to terror when he saw small piles of bones scattered around the camp, and he was sure they were going to eat him. Later in the day, with most of the creatures away from camp, he made his bid to escape, running all the way back to the river mouth where his canoe was stashed. He paddled straight back to Nootka where he arrived nearly frozen and completely exhausted. Father Anthony nursed him back to health, but after his terrifying experience Muchalat Harry never went back into the woods, indeed never left the village again. It is easy to dismiss this vivid account as a complete fantasy in the imagination of Muchalat Harry. He was alone and there were no other witnesses, but something scared him enough to turn his hair white overnight.

Among many other Bigfoot stories from the last century and before, there is another one that I want to tell you about. It has none of the uncertainties of Muchalat Harry's account of his lonely abduction. On this occasion there were plenty of witnesses and a creature was captured alive. It happened in July 1884 about twenty miles north of the small town of Yale, British Columbia

on the banks of the Fraser River. According to the local newspaper, the *Daily Colonist*, a construction crew was heading to
work to continue digging a tunnel through a series of rock bluffs.
Ahead of them, on the track-way, they came across a creature,
which they at first took to be a man, asleep close to the rails.
Woken by the arrival of the train crew, the creature apparently
stood up and dashed up the bluff at the side of the track. Four
of the crew jumped down from the wagon and chased the creature up the rock face, where he was cornered and knocked
unconscious. He was then tied up, lowered to the ground and
taken back to Yale. The *Daily Colonist* goes on to describe his
appearance in detail:

> Jacko, as the creature has been called by his captors, is something
> of a gorilla type standing about four feet seven inches in height
> and weighing 127 pounds. He has long, black, strong hair and
> resembles a human being with one exception, his entire body,
> excepting his hands (or paws) are covered in glossy hair about
> one inch long. His forearm is much longer than a man's and he
> possesses extraordinary strength as he will take hold of a stick
> and break it by wrenching or twisting it, which no man could
> break in the same way.

Then the story goes cold. According to some sources, Jacko was
taken by his captors to England to be become a 'curiosity' but
died on the voyage and his body was thrown overboard. Other
accounts have him exhibited by the showman PT Barnum as
'Jo-Jo the dog-faced boy'. As I have repeatedly discovered, Bigfoot
stories rarely have a clear-cut ending, and this one is no exception. Nevertheless there are few accounts that include the actual
capture of a Bigfoot. Like so many others, the story of Jacko
could be an elaborate hoax perpetrated by the *Daily Colonist*, as
many believe it to have been.

There is another story that deserves a mention. Like Muchalat
Harry's tale it concerns an abduction, but in this case the witness,
far from being terrified by the encounter, spent several days with

his captors and had plenty of time to observe their appearance and their behaviour. The victim, if you like, was one Albert Ostman and the incident took place in 1924. Ostman was a thirty-four-year-old lumberjack working the woods of western British Columbia who decided to have a short vacation and fit in a little gold-prospecting. He had heard stories of a lost mine at the head of the Tuba Inlet, one of the many fingers of sea that reach deep into the mountains north of Vancouver.

After reaching the head of the inlet with an Indian guide, Ostman set off alone. Two days later he found a flat piece of ground, clear of trees and with good views back down the inlet to the sea. He decided this was to be his base for the next few days and he set up camp. His pack was disturbed during the first night, but Ostman didn't think much of it. During the second night his pack, which he had hung from a tree, was emptied out and some supplies, including prunes and flour, were missing. This was more of a puzzle as the two main suspects from the night before, a porcupine or a bear, could either not have reached the pack or would have made much more of a mess of it.

On the third night Ostman slept in his clothes with his rifle close by, intent on catching the intruder in the act. At some point during the night he was picked up, then carried and dragged for about three hours before being dumped on the ground. At first he couldn't see anything, but heard creatures, whatever they were, chattering like monkeys. As the sky lightened with the approaching dawn, he found he was confronted by four large creatures. Once he could see them properly, he reckoned they were a family comprising a very large and muscular male about eight feet tall and around seven hundred pounds, a slightly smaller female and two youngsters around three hundred pounds. Each had exceptionally long forearms with large hands but small fingers and a covering of red-brown hair.

It is not clear how his kidnappers restrained him but after a couple of days he was allowed to wander around their living area. He found their sleeping quarters in a cave excavated beneath a large fir tree and could see that inside the floor was covered by bedding made from interwoven strips of bark and ferns. Over a period of a week, he came to befriend the youngsters and even the large male showed an interest in his possessions, among which was a large tin of snuff. The big male grabbed this from Ostman and swallowed the lot. Not surprisingly, this induced a violent fit of coughing and spluttering. The creature rolled around in agony, his eyes streaming from the effects of the overdose of snuff. Ostman used this opportunity to make good his escape and, after wandering for several days, finally found a logging camp from where he eventually reached Vancouver.

This is only a brief account of Albert Ostman's experiences. He made a far more detailed report that he subsequently swore on oath as a true testament in front of a magistrate. It was the magistrate's professional opinion that Ostman was in full possession of his faculties and certainly believed the story himself, though the magistrate did not commit his own opinion to paper.

One last, very famous case, concerns a violent night-time attack by a group of 'apemen' on a party of miners in July 1924. The miners were in their cabin deep down in a gorge on the northeast shoulder of Mount Saint Helens, a little north of the Oregon–Washington boundary. According to the account of one of the miners, Fred Beck, the creatures attacked the cabin with rocks and tried to break in. The miners, Beck tells us, fought back and fired at their attackers, killing at least one of them, though they did not recover a body. Ever since, the site of their terrifying encounter has been called 'Ape Canyon'.

Though a famous case, the Ape Canyon attack is not free from controversy. In 1983 William Halliday from the Western Speleological Society wrote that it was customary in the 1920s for groups of youths from a nearby YMCA camp to go to the edge of the canyon and throw down the volcanic rocks that were scattered on the surface. Halliday's hypothesis is that on the night in question the campers threw their rocks into the canyon unaware that anyone was below. The frightened miners believed they were under attack and, looking up, saw only the moonlit silhouettes of their assailants, which they interpreted not as the youths but as 'apemen'.

I could fill several more chapters with yeti and Bigfoot stories, but that is not my purpose in *The Nature of the Beast*. There are already plenty of books that do precisely that. I am merely making the case that there is a genuine enigma. How much notice to take of Muchalat Harry or Albert Ostman or the *Daily Colonist* is, as far as I am concerned, unimportant. I would certainly be wasting my time if I were sure that all accounts were patently false or complete fantasy. They well might be, but on the other hand some may be true. And, that in a nutshell, is the enigma.

I must mention the circumstances surrounding the only moving image of a Bigfoot that many believe is genuine. We are once again in northern California. It is 1968 and two friends, Roger Patterson and Bob Gimlin, decide, in their own words, to 'go find a Bigfoot'. Their decision was triggered by reading an article by Ivan Sanderson, who we will meet later, and a reconnaissance trip that Gimlin made to Willow Creek. Patterson and Gimlin arrived with two pack-horses, camping gear and two weeks' supplies, as well as the all-important movie camera. Their *modus operandi* was to wait until the logging trucks had left, then ride up and down the tracks and stream beds looking for footprints. Around midday on 20 October, they left the camp together and rode in the direction of Bluff Creek and then started to go up the right-hand bank, Patterson first, closely followed by Gimlin. After several miles they came across a large fallen tree that had dammed the creek and a pile of logs washed down from a heavy flood four years earlier. Suddenly, from behind the log jam at the side of the creek, a creature stood up. It looked straight at them. Then all hell let loose and the horses panicked. Luckily both Patterson and Gimlin had been rodeo riders and stayed in the saddle. Demonstrating admirable presence of mind, not to mention extreme horsemanship, with one hand on the reins

Patterson reached into his saddlebag, pulled out his movie camera and slid off his horse. The horse ran off and Patterson ran after the Bigfoot. Most people I have interviewed who have seen a Bigfoot actually run in the opposite direction. But not Patterson. Falling to his knees on a sandbar, he raised the viewfinder to his eye and flipped the button to start the film rolling.

The Bigfoot moved quickly, though without appearing to hurry, away from Patterson towards the edge of the clearing, looked back twice then melted into the forest. Gimlin, who had a better view of the creature than the cameraman Patterson put it at around 6′6″ tall and 250 to 300 pounds. At its closest approach, Gimlin was only sixty feet from the creature. He had his rifle in his hand, but did not raise it. It was only there for self-defence and at no time did either man feel threatened. Once the creature had moved into the forest, the two men examined and photographed the tracks it had left in the dust of the dried-up creek.

I don't suppose any piece of film has been so thoroughly examined as the 53 seconds of the Patterson-Gimlin footage. Each frame has been pored over with infinite care, trying on the one hand to learn about the creature and on the other to look for signs of fraud. Athletes of the same height as the creature were taken to the scene and walked the same route. Experts on human locomotion studied the way the body moved up and down. Frame 352, where the creature looks back over its shoulder, received particular attention and is immediately recognisable as the iconic Bigfoot portrait.

A few reputable scientists did take a look at the film soon after it was shot, but most kept well away, fearful of being dragged into perpetuating a hoax. Of those that examined the footage the most prominent was primatologist and expert on locomotion, the Englishman John Napier, then at the Smithsonian Institution in Washington. He did give the film serious attention and arranged for a showing to colleagues at the Institution, most of whom were unimpressed. Although in no doubt that Bigfoot did exist, Napier nonetheless had severe misgivings about the Patterson-Gimlin film and concluded that it was probably, but not definitely, a man dressed in a gorilla costume. Bernard Heuvelmans shared this opinion. Hollywood special effects wizard Stan Winston was less circumspect than the ever-polite Napier. 'It's a guy in a bad fur suit, sorry!' Bob Heironimus, a native Yakima Indian, even claims to have been the man wearing it.

However, Dr Jeff Meldrum from Idaho State University, who as we know from the last chapter has made the study of Bigfoot his principal research area, thinks the film is genuine, as does Peter Byrne. Even I, who have nothing useful to say about it, have been asked for my opinion, such is the enduring ambiguity of the film. It is another part of the enigma, and an example of the confusion surrounding pretty well all Bigfoot claims – confusion I was hoping to cut through wielding the scalpel of genetics.

There are thousands of recorded Bigfoot prints. Some are faked, such as the original tracks found by Jerry Crew at Bluff Creek in 1958. In 2002, following the death of Crew's supervisor Ray Wallace, his son announced that his father had admitted to making the tracks. Crude wooden models of large feet have been constructed in the past but the prints are easily dismissed as hoaxes by experts since they do not even begin to replicate the complex flexions in the walking human foot. Just take a look at your own footprints in wet sand on the beach. The first part of the foot to touch the ground is the outside of the heel, which is why that is the part of a shoe that wears out fastest. Then the contact moves across to the other side of the foot, then back to the ball and finally kicks off from the big toe. The other toes merely prevent slipping backwards, so leave only light impressions compared to the heavy print of the big toe. The rest of the foot, outside the contact zone, leaves hardly any mark at all. Most fakes never approach this degree of sophistication. With others either the hoaxers are really masterful, or the prints are real.

One set of convincing prints was found near Bossburg on the banks of the Columbia River where it bends behind the Cascade

Mountains in eastern Washington State. Like Slick, Oklahoma, Bossburg is now a ghost town, its lead and silver mines exhausted. In December 1969 René Dahinden and John Green, both well-known Bigfoot hunters, found tracks, thousands of them, in snow by the side of the road. What distinguished these tracks was that the right foot was clearly deformed. Although the authenticity of the tracks of the Bossburg Cripple, as the Bigfoot was labelled, was confused by the later claim by a local man Ray Pickens that he had made the feet out of wood and nailed them to his boots, the find was significant enough for the primatologist John Napier to investigate. He did so because he could not accept that a faker would have had the skill or the anatomical knowledge to create a convincing model of a known deformity of the right foot together with an authentic left foot whose print showed the signs of compensation for its crippled partner. These footmarks certainly did not match the crude wooden flippers that Pickens later displayed on television. Another hoax within a hoax? Who knows? But Napier thought they were probably genuine as did another authority, Grover Krantz from Washington State University and, in a recent review of the prints, Dr Jeff Meldrum from Idaho State.

It is very hard to believe that all Bigfoot prints are faked. Witnesses report tracks in snow in very remote places that may never even have been visited before. It really is difficult to accept that in each case hoaxers have taken the trouble to plant fake prints in these remote locations before the witnesses arrived.

Mistaken identity is certainly an issue, with bear tracks the most likely to confuse, especially when the rear foot is placed over the impression of the front foot. Sometimes claw marks are not there, either because the claws are worn down or are held clear of the ground. But such phenomena are well known to experts like Meldrum. When the documentary film crew that followed my own research managed to encourage Brutus, a captive grizzly, to make one of these back-foot/front-foot super-impositions and showed the cast they had made to Dr Meldrum, he immediately identified it as exactly what it was. A bear print.

There are other features of Bigfoot encounters that run consistently through eyewitness reports. One is rock throwing. Large stones often land alarmingly close to frightened witnesses. Many others report that a strong smell alerts them to the presence of a Bigfoot nearby. Some hear wood knocking, as I did myself at the Big Guy's tree, as though a trunk is being struck by a hard object. Several witnesses, both in the Pacific Northwest and the Himalayas, report hearing a mewing call rather like a seagull but louder. For some, branches broken or twisted in certain ways are sure signs that a Bigfoot has passed through. The rare visual sightings also share consistent features. The creatures are usually tall, hairy and very muscular and yet also very wary. In most encounters, the Bigfoot stands up, looks at the astonished witness for a few seconds then slowly walks into the forest gloom and is lost from sight. There are no signs of aggression, and witnesses, though always surprised and usually frightened, do not feel in the least threatened. Many people I talked to felt they are being watched when in a known Bigfoot hotspot. Lonely forests are strange places and it is easy for your imagination to run wild. All the same it really does sound as if there may be 'something out there' after all. And that, of course, is another ingredient of the enigma.

Why has it been so very difficult to get hold of good proof? There is no undisputed photographic evidence, even in the days when almost everyone has a camera phone. There is shaky mobile phone footage, but how much of that is faked, especially these days when Photoshop and other programs are widely available to manipulate any image? I had an interesting conversation with one of the cameramen in the film crew that was making the documentary following my project. He said it would be easy to make a convincing film of a Bigfoot and even explained how he would go about it. He would first film a man in a gorilla costume moving through woods using the highest definition that his equipment could achieve, which was very high indeed. These shots would be taken while ostensibly filming another event altogether – he suggested a cycle race. Then he would degrade the images to create the right degree of 'fuzziness' and finally launch the footage at a press conference, expressing his extreme surprise that he had caught the creature on film entirely by chance. If you see anything like this in the future you will know where to look first.

When Peter Byrne arrived in Willow Creek in 1959 after accepting Tom Slick's suggestion that he switch from hunting yetis in the Himalayas to tracking Bigfoot in the Pacific Northwest, he had to change tactics. In the Himalayas, Byrne and other members of the Slick expeditions did most of the tracking themselves, albeit with local assistance. While he had relied on the trustworthy and hard-working Sherpas to help him in Nepal, he soon discovered that the available manpower in Willow Creek was in a very different league from the Himalayas. In northern California the replacements recruited by Slick's agents were a motley crew of loggers and woodsmen. If they were to be believed, the hunt for the Bigfoot was almost over. According to the leader of the 'associates' that Byrne had been assigned, his companions were hot on the trail of a Bigfoot and a live capture was only days away.

And so it seemed when the phone rang shortly afterwards and an excited voice announced that a young Bigfoot had been captured. The creature was covered in hair from top to toe and had the most enormous feet. When Peter enquired when he could inspect the creature the reply came: 'Just as soon as your sponsors come up with the price, that's when.'

'And what would that be?' Byrne asked.

'One million dollars.'

For that they would hand the Bigfoot over, and throw in the cage too. Byrne called Slick who told him to offer five thousand. That didn't work.

Two days later the callers rang back and said they would settle for five thousand dollars, paid in advance. 'No chance,' replied Byrne. A few days later he took another call. The men were getting desperate. The creature would only eat frosted flakes and they were running out of cash for new supplies. Again Byrne contacted Slick who authorised the offer of five hundred dollars for a look at the creature and a photograph. Again the answer was no. A few days later another call informed Byrne that the creature was sickening and, out of the goodness of their hearts, they had released it. Needless to say, Byrne decided against retaining their services any longer.

Byrne, his brother Bryan and a professional hunter, Steve Mattice, often joined by Tom Slick, continued their search for evidence of Bigfoot helped, though sometimes hindered, by numerous reports of Bigfoot encounters from members of the public. During 1960 they found twelve sets of footprints, but did not see a Bigfoot. Just as the Pacific Northwest Expedition, as it was called, was getting properly established, tragedy struck. In October 1962, Slick and his pilot were caught in a violent thunderstorm over Dillon, Montana and their two-seater plane broke up. They were both killed. Slick's various Bigfoot expeditions – there had been another in British Columbia, for example – came to an abrupt halt and all the material that had been collected, like that from the Himalayan ventures, disappeared. Peter Byrne returned to Nepal and resumed his big-game hunting.

With Slick's death the hunt for Bigfoot lost its only wealthy and committed private backer. Though the search was never abandoned, and continues to this day, it was no longer able to rely on the deep pockets of Tom Slick.

Peter Byrne managed to arrange sponsorship through the Academy of Applied Sciences in Boston and returned from Nepal in 1972 to run the Bigfoot Information Centre located in The Dalles, Wasco County, Oregon. From this picturesque setting on the banks of the Columbia River, the Centre systematically recorded Bigfoot encounters from the whole of the Pacific Northwest, sending trained staff members to interview the witnesses and check their accounts. Though some of these witnesses were obviously cranks, most were ordinary people who had had an extraordinary experience. The Centre in The Dalles closed through lack of funds in 1979. Peter was back in Oregon by 1992 running a similar project in nearby Parkdale until 1997, but with no more positive results than at The Dalles. When I visited him at his home in 2013, he was still actively engaged in the hunt for Bigfoot, even at the age of eighty-eight, this time with a team of local people and concentrating on encounters in the nearby coastal range. I met some of his team and listened with eager ears to their plain-spoken accounts.

In my opinion it is the accounts by hundreds of apparently rational and sensible people that build the argument most strongly for one side of what I see as a genuine enigma. These witnesses, who in every other respect are perfectly normal, who have nothing to gain yet have seen, often in good light, a large ape-like creature walking on two legs. To dismiss all of these accounts out of hand seems to me to be just as unscientific as accepting them without further evidence.

Enigmas have two sides and the obverse here is the plain fact that no bodies have ever been found, nor is there any un-ambiguous photographic evidence. The footprint evidence, after elimination of fraud and artefact, certainly begs the question of what made them, but even Dr Jeff Meldrum, who has spent most of his professional career studying these tracks, concedes that this is not proof of Bigfoot's existence. And he is right. No matter how many thousands more plaster casts are made, they will never be accepted as unambiguous proof that Bigfoot exists. For that, something else is required.

7

The Russian *Almasty*

While legends of wildmen flourish all around the world, active searches for anomalous primates have been confined to the yeti in the Himalayas, the Bigfoot/sasquatch in the US and Canada and in one other region – Russia. Yeti-hunts in the Himalayas have been, on the whole, substantial expeditions and sometimes, as in the case of Sir Edmund Hillary's in 1962, coupled to a mountaineering objective. Bigfoot searches in America and Canada are organised quite differently and largely carried out by individuals acting entirely alone or in loosely co-ordinated groups. In Russia, things are different again, and their search for anomalous primates is the only effort to have enjoyed any official government backing.

The instigator and the central figure in Russian hominology, as the science is called, was the late Boris Porchnev. It was Porchnev who was the first to suggest the intellectually attractive hypothesis that the local anomalous primates were, in fact, surviving Neanderthals. There is plenty more to say about my efforts to test Porchnev's theory later in the book, but for now let me concentrate on the abundant sightings of Russia's own yeti, usually called the *almasty*. The *almasty*, which has been seen in many parts of Russia but especially in the Caucasus and in Siberia (where it is also known as the *alma*), shares many of the attributes of both the yeti and Bigfoot: generally tall, though short varieties are also reported, muscular and hairy in appearance, wary and retiring in behaviour.

Porchnev, though an academic historian by training and an authority on the French Revolution, is best remembered for his pioneering work on the *almasty*. It was through his influence that the USSR Academy of Sciences established the 'Snowman Commission' in 1958 to co-ordinate all reports of these creatures and to organise expeditions to find them. This was an impressive achievement. I cannot imagine the US and UK equivalents, the National Academy of Sciences and the Royal Society respectively willing to take such a risk. How differently things might have turned out had they done so.

During its short life, the Snowman Commission sent an expedition to the Pamir Mountains in what is now Tajikistan. It was stimulated to explore this region by a rare *almasty* sighting by one of the Academy's own members, the geologist Alexandr Georgievitch Pronin, who saw one at the edge of the Fedchenko Glacier in August 1957. Standing on a boulder, at first he thought it was a human, although he knew the area to be uninhabited. The creature was stocky and stooped with long, trailing forearms visible as it walked across the snow before disappearing among the rocks. Unusually for any sighting, yeti, Bigfoot or *almasty*, he saw the creature again three days later in the same vicinity. Pronin later reported what he saw to the Snowman Commission with mild surprise, saying, 'I had heard reports of these creatures, but never expected to see one myself.'[1] Nonetheless, the official expedition found nothing when it searched the same area. There being no definite proof of the kind the Academy required to declare that the *almasty* existed, the Snowman Commission was dissolved soon after the expedition returned empty-handed.

Almasty enthusiasts are not easy to kill off and Porchnev found the survivors a home in the Darwin Museum in Moscow by inaugurating a monthly seminar in 1960. Fifty-four years later, this monthly seminar series is still going. In August 2013 I had the privilege of giving one myself.

Porchnev was also in contact with a remarkable Mongolian academic, Yöngsiyebü Rinchen. A linguist by training, Rinchen had made a particular study of Mongolian *almas*, including retrieving and studying several promising skulls. Images of these were published by the Dutch lawyer and anthropologist Tjalling Halbertsma, who has lived in Mongolia for many years, but the skulls themselves have long since disappeared.[2] In Halbertsma's account, Rinchen was a prolific correspondent, something I can certainly verify having seen his many letters to Heuvelmans in Lausanne. Halbertsma also writes that Rinchen sent two hairs to Heuvelmans, but search though I did throughout the Rinchen folders in the archive, I could not find them. Rinchen collected stories of *alma* encounters from the people of remote regions of the country. There was even a body, covered in hair, found in a gorge. The skull and some bones were brought back to the capital, Ulan Bator, where Rinchen worked, but again they too have vanished.

Despite the disbanding of the Snowman Commission after the unsuccessful expedition to the Pamirs, the search for the *almasty* continued thanks to the efforts of individuals, both academic and lay. Porchnev was a close friend and colleague of Bernard Heuvelmans, whose archive in Lausanne contains especially plump folders of their correspondence. Together they wrote *Les Neanderthals Sont Toujours Vivant*, a book that amplifies Porchnev's surviving Neanderthal theory with Heuvelmans' account of the Minnesota Iceman, of which more later. Though Porchnev died in 1972, his legacy has been continued by his protégé Igor Burtsev and his colleagues Dmitiri Bayanov and Michael Trachtengerts. I was delighted to find that this triumvirate of scholars was still active, and eager to talk, when I visited Moscow in the summer of 2013.

8

The Godfather

Before long, anyone who reads about the yeti is led to the work of Dr Bernard Heuvelmans. Heuvelmans was a qualified biologist, born in 1916, who spent his working life gathering scientific evidence for the existence not only of yetis and associated wildmen, but all sorts of other creatures 'unknown to science'. His 1955 book, *Sur la Piste des Bêtes Ignorées*, translated into English two years later as *On the Track of Unknown Animals*, rapidly became a bestseller. It is still in print. After its publication Heuvelmans was widely acclaimed as the undisputed leader in the newly named field of cryptozoology. Very fortunately, one of my senior colleagues in Oxford remembered being given a copy of Heuvelmans' book as a boy soon after it had been published. The next day I held a faded first edition of the cryptozoology equivalent of the Old Testament in my hands.

Heuvelmans died in 2001, aged eighty-four, and it did not take me long to discover that he had bequeathed his entire archive to the Museum of Zoology in Lausanne, Switzerland. A week later, I was climbing the wide stone steps that led up to the entrance of the Palais de Ruminé, Lausanne, in whose splendid interior is housed the zoology museum. I knew within minutes of meeting the museum director, Michel Sartori, that he and I shared many of the same opinions, even suspicions, about cryptozoology. Michel is a biologist, a specialist in mayflies, and although our fields are a long way apart we both have the same philosophical outlook on science, the belief in the crucial importance of evidence that binds all professional scientists. In fact, it was the philosophy of the project that soon became the focal point of our discussion.

I had been struck by a persistent thread running through Heuvelmans' book, that cryptozoologists feel strongly that they have been shunned by the 'scientific establishment', or even by 'science' itself. I was soon to discover why Heuvelmans himself felt that he had been ostracised by the establishment. Both Michel and I completely disagreed with the notion that science can simply reject the claims of cryptozoologists without a proper examination of the evidence. As the curator of the Heuvelmans archive, Michel has a duty to make it available to genuine enquiry. In the ten years since the museum acquired the archive there have been a number of visitors from the world of cryptozoology but, until I turned up, none from professional scientists. Both Michel and I were both very well aware that cryptozoology has more than its fair share of cranks and hoaxers but we both felt that this complication did not exclude the subject from the realm of proper scientific investigation.

After an hour, Michel led me to another part of the building and swung open the heavy metal door of a strongroom containing the Heuvelmans archive. Inside were rack upon rack of metal shelving mounted on rollers to save space. We opened up the second tier and there on the shelves were about a hundred box files, their spines illustrated with photographs or drawings. I had seen images of many of these unique treasures before, in an old photograph of Heuvelmans at work in his study, so to be able to lift one off the shelf and open it was a thrilling experience. Each of the boxes was filled with carefully ordered copies of press cuttings. Every conceivable mention of creatures 'unknown to science' that had been published anywhere in the world between the late 1950s and 2001, the year of his death, was here, largely undisturbed for the last decade or so. Beneath the box files, suspended in folders arranged in alphabetical order, was the entire correspondence of his last fifty years.

Many of the Heuvelmans files were decorated by photographs or drawings of young women in erotic poses. Women, I realised, must have played a central role in Heuvelmans' life. Many images were of his former wife, Monique Watteau, a painter whose short marriage to Heuvelmans was only the start of their relationship. After their divorce she went on to marry Scott Lindbergh, son of the aviator Charles, whose solo transatlantic flight in 1927 secured his enduring celebrity. The newly married Lindberghs moved to a chateau in the Dordogne, and took the impoverished Heuvelmans with them to live in a small house in the grounds. Here Monique Lindbergh, as she now was, worked on the illustrations for Heuvelmans' books, including *Sur La Piste*. After a long affair with the Hollywood star Yul Brynner, she changed her name again to Alika, her *nom d'amour*. Alika Lindbergh remained close to Heuvelmans, who himself enjoyed a string of girlfriends throughout his long life.

I wanted to look at a sample of Heuvelmans' correspondence, so I lifted out a suspension file in which his letters had been preserved between thick layers of acid-free paper to prevent fading. As I had a particular interest in his correspondence with the Mongolian scholar Yöngsiyebü Rinchen, I chose the folder marked 'R'. At the front of the file was a small sheet of writing paper embossed with an extravagant crest. It was from Prince Rainier of Monaco, himself a keen marine biologist, thanking Heuvelmans for a copy of his latest work on sea monsters. As I was to discover, Heuvelmans was extremely well connected, and as I leafed through the cuttings and the correspondence stored in this remarkable archive it was obvious that he had spared no effort in trying to find the extraordinary proofs for the existence of his creatures that science and the 'scientific establishment' demanded. In the end he was disappointed. Neither his creatures, nor cryptozoology, the field he inspired, were ever accepted during his lifetime.

I could not help wondering what might have happened had Heuvelmans been able to use the new techniques of genetics that were now at my disposal. I was sure he would have welcomed them, and seen their potential to bring cryptozoology into the scientific mainstream. It was as if I had now been given the tools to finish the job that Heuvelmans had started, and I began to formulate my plans around that tantalising thought. I had never met the man, and am not at all sure I would have liked him had I done so. But I was confident that, if these creatures really were new species and if I could get hold of physical remains with some DNA in them, then I could prove their existence in a way that Heuvelmans never could.

Descending the wide stone steps of the museum to the Place de Ruminé, I found a street market in full swing. I was feeling very pleased after my meeting with Michel and excited by my glimpse into the Heuvelmans archive. Some new projects are hard going from the beginning, but with this one doors were swinging open with the lightest touch, which is always a good sign. Browsing through the market stalls, I came to one selling second-hand books. There on the table was a copy of *Tintin au Tibet* with, on the front cover, a trail of giant footprints leading up a snowfield to the peaks beyond and looking just like those in the Shipton photograph. This was too promising a discovery to ignore, so I bought the book. Of course, during their adventure, Tintin, Snowy and Captain Haddock encounter a yeti. But that was not the end of the coincidences.

I later learned that Georges Remi the cartoonist who created Tintin, under the *nom de plume* Hergé, was a close friend of Heuvelmans, a fellow Belgian. It was Heuvelmans who had suggested Hergé model the head of his cartoon yeti on a scalp Heuvelmans had collected from Nepal a few years earlier, and which I also discovered on my next visit to Lausanne was on display in the museum – and available for DNA testing. With luck smiling so sweetly, how could anything possibly go wrong? A few days later, Michel and I agreed to make the yeti project an official collaboration between our two institutions, and the Oxford-Lausanne Collateral Hominid Project was born.

Under the eaves of the Palais de Ruminé is a nesting colony of the rare Alpine swift, larger than the more familiar common swift whose piercing cries in the skies above Oxford are the heralds of high summer. In Lausanne, the Heuvelmans archive is on the top floor of the museum and through the open window I could hear the chattering of the swifts as they pursued each other around the square and dived into the narrow crevices that hid their nests.

The same sound welcomed me back to the archive the following year. By then the project was well under way, and samples were arriving for DNA analysis from all over the world. I had put a week aside for the archive but without any particular objective in mind. By now I had read enough to know who Heuvelmans' main correspondents were likely to be: Boris Porchnev, Yöngsiyebü Rinchen, Peter Byrne perhaps and certainly his close collaborator Ivan Sanderson, who we shall meet very soon. And John Napier, William Osman Hill and many more that we have yet to encounter. I began by looking at the letters Heuvelmans had received from these men – yes, almost all cryptozoologists have been men. There were also file after file of press cuttings, many sent from agencies to which Heuvelmans had delegated the tedious task of spotting relevant articles in newspapers and magazines from all round the world. Other files overflowed with details of cases that Heuvelmans had investigated himself.

One in particular caught my eye. Its spine was decorated with a photograph of the dead body of a strange ape-like crea-ture, its limbs contorted into a very unnatural pose. Inside I uncovered letters and documents relating to the case of the Minnesota Iceman. It was a very famous case for cryptozoolo-gists, but as I leafed through the material, the relevance of this episode for Heuvelmans himself became more and more clear. On the desk in front of me were the original letters that documented the excitement, the betrayal and the despair that finally destroyed Heuvelmans and severed the link between cryptozoology and mainstream science for the next four decades. Such is its significance, that I have no hesitation in devoting the bulk of this chapter to the enthralling Case of the Minnesota Iceman.

The story begins in December 1968 when Heuvelmans was on his first visit to the US. After more than a decade of correspondence he had finally met up with Ivan Sanderson. Sanderson was another of the colourful characters who decorate the history of cryptozoology. Scottish by birth, Sanderson had studied natural sciences at the University of Cambridge. Then, after graduation, . he worked in counter-intelligence, had spent several years as a taxidermist, then as a publicist and an author of books on cryptozoology. His letters to Heuvelmans are typed on paper embossed with his affiliation. Not Yale or Harvard but 'The Society for the Investigation of the Unexplained', based in his home in Columbia, New Jersey.

Earlier in the same year Sanderson was trying hard to persuade Heuvelmans to emigrate to the US. He writes, with a lot of underlined accentuations that I have retained:

Mon Cher,

Your letter of the 23rd Jan reached me only today. I am horrified. I can see by your hand-writing that you are 'in the dumps'. Bernard, you've just GOT to climb out of it yourself. You have no friends – none of us do – they are all self-centred leeches – and there is nothing wrong with that – it's just part of the biological process. BUT . . .

I've been batted down more times than I can count. Recently I really 'had it' with all the full forces of all the 'establishments' and a lot of undeniable others besides; but, as of now, I am still here. Don't 'give up the ship': but first get your health straightened out. And this time, take my advice.

After that, I being completely and absolutely, and fucking-well bankrupt, can not send you any cash – which, believe thou me, I <u>would</u> do if I had any.

Point is, why in the effing-hell don't you just 'up-stakes' and come over here on a tourist or visitor's visa. I'll try to get up the cost of your passage (with return, to satisfy the immigration). <u>We'll</u> look after you somehow: and, as I have said before, if you do ever haul-arse over here, I'm prepared to bet that you will be a full professor at some topnotch university within three months.

Your friend, Ivan.

This characteristically feisty appeal from Sanderson touched on two pressing issues for Heuvelmans. First his chronic lack of cash. Despite the runaway success of his first book, which brought in a small fortune in royalties, money slipped through Heuvelmans' fingers like water. He had numerous 'female friends', spending two months every summer in their company on an island off the Côte d'Azur. As if that were not enough of a drain on his finances, he was always helping out friends or acquaintances, and even strangers, who fell on hard times. As the Sanderson letter hints, Heuvelmans was by now dependent on handouts from his own coterie of disciples.

The second sensitivity that this letter reveals was Heuvelmans' failure to secure a proper academic position after getting his PhD. Until *Sur La Piste des Bêtes Ignorées* appeared in 1955, he supported himself by writing newspaper and magazine articles on a wide range of topics, many with a technical or scientific focus. Among many preserved in the Lausanne archive are essays on subjects as diverse as Leo Baekeland, the inventor of the early plastic Bakelite, on the impact of the newly arrived medium, television, on 'global cooling' following a warning by two American scientists that 'soon there will be no Summer', on the possibility of Martian invasions and on the Swiss explorer Jacques Piccard's descent by bathysphere to the ocean depths. These articles were written either under his own name or using a pseudonym, Dr Simon Obispo. He also wrote about sex, either as Dr B Heuvelmans, as in the scientific treatment 'Sex Appeal' or, more commonly, in erotic fantasies such as 'Ménage à Trois' or 'The Naked Hitchhiker' where his authorship was thinly disguised as Barney Hillman (the English translation of Heuvelmans).

On 9 December 1969 Heuvelmans was staying with Sanderson at his home in New Jersey when a telephone call arrived from a Mr Terry Cullen, the owner of a reptile vivarium in Milwaukee, Wisconsin. Heuvelmans' handwritten note shows the following text as having been taken from the shorthand transcript by Marion Fawcett, Sanderson's secretary, immediately after the call.

An American is now touring local fairs with the body of an alleged ABSM (*the shorthand for 'Abominable Snowman' used throughout the archive*) frozen in a block of ice. This is reported to have been obtained from the Red Chinese who stole it from the Russians. The American got it from the Chinese in Hong Kong harbor.

It was apparently found by a Russian fishing trawler which thought it was a seal frozen in the ice. As the ice melted somewhat, a 'monkeylike' form became visible.

Mr Cullen has seen this and describes it as a 'rather hairy hominid'. It has a sagittal crest (*a prominent bony ridge on top of the skull used to anchor powerful jaw muscles*) but no canines. The back of the head has been smashed and the brains are 'hanging out'. The feet are human, the great toe not being opposed. The hair is dark brown and 3–4 inches long; Cullen states that the hair grows out of the 'pores' and is not pasted on. The body is apparently somewhat sprawled out but he estimates its height at 5 to 5 1/2 feet and emphasizes a barrel chest and wide shoulders.

Soon after the news reached Sanderson, he set about finding corroboration of Cullen's story from his network of members of the Society for the Investigation of the Unexplained. Confirmation soon arrived from Member No. 215, a Mr Richard Crowe of Chicago, who had inspected the specimen when it was on display in his home city. Sanderson also records reports from 'several dozen' people who had seen the exhibit in different locations around the Midwest.

Imagine how thrilled Heuvelmans and Sanderson must have been to have a real body to examine. Then, much as now, many commentators had asked why, if these creatures existed, no body had ever been obtained for proper scientific examination. Here was the opportunity for the final vindication, a chance to silence the critics once and for all, and to make what would be the greatest scientific discovery of the twentieth century. All this without the need to mount an expedition to some far-flung corner of the earth. The proof was in a freezer, somewhere in the American Midwest. Almost on the doorstep.

Through his network of contacts, Sanderson quickly located the specimen at the ranch of Mr Frank D Hansen, near Winona, Minnesota and, on 14 December 1968, he and Heuvelmans set out by car from New Jersey. Before they left Sanderson noted:

> The crux of the matter is that we have now a (fresh) corpse of at least one type of ultra-primitive hominid, with fresh blood, available for proper examination.

Driving through the Midwest in winter is no joke, but they arrived at the Hansen ranch on the morning of 17 December. The ranch was located on top of the plateau that lies to the west of the Mississippi and could only be reached via a bewildering complex of unsigned dirt roads. Nevertheless they got there and were soon gazing at the creature lying in a large chest-freezer that was kept inside a trailer in the yard. There was no doubt that this was a body of something very strange, frozen beneath a translucent layer of ice.

The pair spent four hours that afternoon examining the object, returning the following day for a further four hours of photography. The weather was getting much colder, and Hansen invited them to stay the night. Next morning, 19 December, they spent three more hours making scale drawings of the body. By then the pair were feeling very cold inside the trailer and Hansen tried to make them more comfortable by installing a gas burner. By the time they left that afternoon, Heuvelmans and Sanderson were beginning to feel the disorienting effects of carbon monoxide poisoning and put up in the first motel they could find. They returned to Sanderson's home in New Jersey on 22 December after eight days and 2,600 miles on the road. The two men had spent a total of eleven hours studying the specimen and they were left in no doubt that it was genuine.

Frank D. Hansen had gone into the carnival business in the early 1960s when he had exhibited a restored 1916 classic four-cylinder John Deere gasoline tractor, one of the very first to have been manufactured. He had refurbished an old creamery on the ranch to work on the tractor and his other exhibits and, after the Iceman, was planning to tour with a show about UFOs. Hansen told Sanderson that he had been an army flyer for seventeen years, mostly in the Far East and that, like Sanderson, he had been engaged in counter-intelligence. He and his wife Irene had been married for twenty-six years with a son and a daughter living locally. Only Hansen was permitted to drive the trailer containing the Iceman, though his wife always accompanied him when they were on the carny circuit. Of course, Heuvelmans and Sanderson wanted to know how Hansen had acquired the Iceman and the notes record four different accounts of its history, accounts that Sanderson prefaces by the following two paragraphs:

The exhibit is alleged to have been on the road for no less than eighteen months. Whether this is true or not is of little import since it was definitely at the famous 'Stock Show' in Chicago in November, 1968, where it was seen by our first informant (*Terry Cullen*). Further, that it was there was confirmed to our member No. 215 by the lady who acts as executive secretary of the organisation that runs this show annually. Further, and subsequent to our inspection visit, several other people told us that they had seen it in Milwaukee and other cities – some even saying <u>more</u> than eighteen months ago.

The other aspect of the history is the 'origin' of the object. To this there are four versions, the first two still not confirmed by Mr Hansen directly.

(1) The story as relayed to us by Mr Cullen, who said he got it from Hansen at the Chicago Fair, was that it had been retrieved from the sea off Kamchatka by a Russian sealing vessel, the captain of which thought it was some animal. The block of ice lifted aboard weighed some 6,000 pounds. This sealer had to put in at a Red Chinese port in an emergency and was virtually seized by the authorities. After a near fight, the ship and crew were released, but after they were at sea, the mate reported that the Object, along with the rest of the cargo, had been off-loaded by the Chinese. The Object is said to have disappeared in China for 'several months' but finally turned up in Hong Kong as contraband. By this version, Hansen was led to it there and paid a lot of money for it.

(2) The second version that Hansen gave a couple of weeks later was that it had been a Japanese whaler which had obtained it and that its owners had sold it to a Chinese curiosity dealer in Hong Kong. When asked about this, Hansen remarked casually that . . . 'Oh that whole thing's a mess. We don't know the name of the ship and can't find it.' Pressed, but for diplomatic reasons not too strongly, he went on to say that it had been found in Hong Kong by an American film executive on a trip to the Orient for 'background material' and that he (Hansen) had then been sent to fetch it with the necessary money. This brought up the mysterious (and I may say, to me, at least, somewhat dubious) Mr 'X' of Hollywood, of whom more anon.

(3) Having been alerted by Terry Cullen by phone that Hansen was extremely wary, had changed his story and was scared almost into silence by 'the government', I tried to set a friendly and co-operative stage at the outset of our first meeting by remarking casually, in Heuvelman's presence and that of Hansen's brother-in-law, Walter, that, 'Seals are constantly found embedded in flow-ice in international waters in the Bering Sea.' Hansen pounced on this, remarking that he never knew it, but looked very relieved. (Note: this will probably be his official story from now on).

However, Sanderson's fourth paragraph adds a further twist to this already tangled history.

(4) During the last hour of my inspection of the corpse, I noted something that ordinary lights had not disclosed. I had powerful floodlights to one side, the beam from which shone <u>under</u> a layer of opaque ice covering the center and right side of the torso. Almost in the middle of the torso there is a low dome of fresh, dark opaque ice which was explained to us by Hansen as follows. He alleged that when the top layers of ice were being shaved away by the professional ice-carver in Los Angeles, he went too deep and uncovered a 'nipple' which immediately turned black and showed signs of rotting. He therefore re-gelated over it. However, I immediately spotted <u>both</u> true nipples, which were pink and fresh and to either side of the torso just like ours. Under the central ice dome I could see (since I was looking straight down while doing a scale drawing) what looked like a bullet hole around which was a little red blood in the crystal-clear ice.

This last observation, if confirmed, places an entirely different connotation on the whole affair. (The legal implications are discussed below). It adds a fourth possibility: namely, that the creature was shot <u>in this country</u> and,

after some medical man had pronounced it to be more
hominid than pongid (*i.e. more man than ape*), it was
quickly and 'obviously' 'buried' (in ice) and put on the
carny circuit as the best possible 'cover'.

The implied possibility of homicide had clearly alarmed Hansen,
as it had the mysterious 'owner' who had been 'hopping mad'
that anyone had been allowed to inspect the corpse. Alerted to
the possibility of an official murder inquiry, or at least a range
of criminal charges surrounding the possession of a human corpse
and taking it across state lines, Hansen and the Iceman disappeared for a while. Sanderson and particularly Heuvelmans were
very worried that the corpse would be destroyed and their chance
of a proper examination would be lost forever.

This is what they saw, again taken from Sanderson's own account:

The specimen is preserved in clear ice in an insulated coffin
with quadruple glass top, and four strip lights shining inwards
and downwards. The coffin measures internally seven feet three
inches by three feet. The insulation is some form of foamite
devised for arctic conditions. Externally it is made of veneered
plywood of double thickness, and there is a chromium-plated
guard rail around the top that visitors can lean on. It occupies
the whole trailer but for a small sleeping bunk, cooking stove
and toilet, forward. The whole coffin unit weighs just over 4,000
pounds and had to be inserted with a heavy 'lift'. The right hand
side of the truck opens longitudinally at waist level, the top
going up and out as an awning; the bottom falling to the ground
to carry a large 'come-on' sign.

It was explained to us by Mr Hansen that the block of ice
containing the specimen was initially over nine feet long, five
feet wide, and over four feet deep. They had a professional ice-
carver shave off about two feet on top, going as deep as possible
without breaking through onto the specimen. This results in a
sort of 'mountainous' surface in low relief, the domes of ice
going up and over the feet, knees, torso and head.

The ice above the body is crystal clear but for a large number of patches of what are obviously hair-fine tubules through which gases are forcing their way out of the corpse and causing 'pencils' of white opaque crystals. These are particularly noticeable exuding from the nostrils, the mouth, a break in the left forearm, the hole in the chest, and the groin. Further, the right-hand side of the body is almost entirely hidden under opaque but white crystalline ice.

So much for the circumstances of the corpse, written in the journalistic prose style of Ivan Sanderson. For the detailed description of the corpse itself, I am turning to the more prosaic assessment of Bernard Heuvelmans. Though very capable of dramatic description, remember that Heuvelmans was writing on this occasion for an audience of established scientists, among whose ranks he was so keen to be counted. This was his chance of acceptance, so suppressing the excitement he no doubt felt, he gives this deliberately sober account of the kind that would win him this acceptance, and much more. Perhaps even the Nobel Prize.

Under the heading 'General Description' he begins:

The specimen at first sight looks like a man – or if you prefer an adult human of the male sex – of rather normal height (6ft) (*Sanderson put his height at between 5' and 5' 6"*) and proportions, but excessively hairy. It is entirely covered with very dark brown hair, 3 to 4 inches long. Its skin appears wax-like, similar in colour to the cadavers of white men not tanned by the sun. One can easily see this on all the naked surfaces of the body – and also on the middle of the chest and on the knees – the hair roots being generally more than 1/16 inch apart.

The left arm is twisted behind the head with the palm of the hand upward. The arm makes a strange curve, as if it were that of a sawdust doll, but this curvature is due to an open fracture midway between the wrist and the elbow where one can distinguish the broken ulna in a gaping wound.

The right arm is twisted and held tightly against the flank, with the hand spread palm down over the right side of the abdomen. Between the ring finger and the medius, the penis is visible, lying obliquely on the groin. The testicles are vaguely distinguishable at the juncture of the thighs.

The head is tipped back so that the mouth and the bottom of the nose are the highest points. Despite this position, which in a modern Man would free the neck and possibly make the Adam's apple jut out, the junction of the neck and torso, hidden deep within the ice, is not visible. This leads me to suspect that this specimen has a very short neck, or, at least, normally carries its head low on the shoulders. (Unless it has been beheaded).

Besides the fracture of the forearm already mentioned, the specimen has several other visible wounds. The head is the most seriously damaged. According to Mr Hansen, who was able to see the dorsal side of the specimen before the block of ice was placed in its present container, the entire occipital part of the cranium was broken or knocked out, and some brain material was hanging out. In the depths of the ice, large smears of blood are visible in this region. The right orbit is empty and bloody. The left eyeball is out of its socket and rests on the cheekbone.

The splintering of the occiput (*the back of the skull*), and the fact that the eyes are out of their sockets, suggests that the specimen may have been struck by bullets from the front. It may be that a bullet struck the forearm as the creature attempted to protect himself. A second bullet then probably entered the right eye, destroying this, and by its impact forced the other eye out of its socket and blew out the rear of the skull, which would have led to its immediate death.

Heuvelmans goes on to explain his reasons for believing that the Iceman is a new hominid species.

Besides the excessive hairiness and the apparent shortness of the neck, a minute examination reveals many traits incompatible with the anatomical characteristics of a normal individual belonging to one of the five known races or sub-species of modern Man (*Homo sapiens*), according to the recent classification of Carleton S. Coon (1962). (*Coon was the pre-eminent anthropologist of his day, although his system of racial classification has long been abandoned.*)

The hands and feet appear to be of strikingly abnormal size and thickness. The fingers and toes seem as large and massive as those of an adult gorilla. It may be that the ice, acting as a lens, exaggerates their size slightly, but this cannot account for the peculiar proportions nor the actual measurements.

The foot has a non-opposable great toe, which is typically hominid. The toes are almost of the same thickness (1–1 1/4 inch across) and of the same length – the 'big' toe is hardly wider or longer than the others. The extremities of the toes seem to be almost all on the same line in the frontal plane; i.e., in respect to the anteroposterior axis of the foot, the line made by the front end of the toes <u>seems</u> perpendicular to this axis and not trailing off obliquely, as in the foot of modern Man. It is not impossible that once the foot is stretched out, digits II and III will extend beyond the great toe, as is seen in the Neanderthalers and in certain abnormal individuals of our own species. The foot gives the impression of being relatively shorter and more thickset than in modern Man, and is eight inches wide across its widest part. The sole of the foot is much more wrinkled and appears to be divided into more pads than in modern Man, thus resembling the great ape's feet.

The hand is wide and relatively short, as in certain modern Men, but much larger: it is 11 inches long and 7 1/2 inches wide. The thumb is extraordinarily long. Extended beside the index finger it would appear to reach <u>at</u> <u>least</u> as far as to the proximal apophysis of the intermediate phalanx (*i.e. to the first joint of the middle finger*) – not somewhere along the proximal phalanx as in modern Man. There are short, heavy nails of a yellowish colour on both fingers and toes.

The thorax appears moderately strong, but is more rounded than in modern Man. Moreover, it joins the abdomen in such a fashion as to form a barrel-shape, twenty-seven inches long. Whereas in modern Man the waist is clearly marked by a narrowing at the height of the navel, in the present specimen a narrowing of the body is only visible at the level of the pubis, at the junction of the muscles of the trunk and thighs, as is the case with anthropoid apes.

The nipples are pinkish and positioned as in modern Man. The navel cannot be seen because of the opacity of the ice and proximity of the thumb. The penis is very slender and about four inches long. The testicles seem to be rather small.

The arms are quite long (forty inches, hands included) and apparently reach to the knees when hanging. The legs appear also to be abnormally long, but this is purely illusory due to the fact that the equally hairy great primates we are accustomed to seeing – the anthropoid apes – have very much shorter legs. In fact, the legs of the present specimen are only 35 inches long from the hip-bone down, but this is within the normal limits of variation seen in modern Man.

Many details of the head are hidden in the depths of the ice. The mouth is almost flush with the surface of the ice, which is partially 'frosted' at that spot. This makes it extremely difficult to see all the details of the straight, widely cleft mouth whose lips – i.e. the everted mucous membrane of the mouth – are extremely fine, indeed almost non-existent. The mouth is slightly open, and one can see a yellowish tooth, which seems to be the right superior canine, but may be the right superior incisor. In any case it is not particularly strong, pointed or long: in brief it is more incisiform like the canine of modern Man.

Two wrinkles mark the skin on both sides of the mouth and perpendicularly to it. The chin appears rounded and receding. One cannot distinguish the ears. One can clearly see the widely flaring and circular nostrils. The nose appears to be short, even retroussé, but not flattened. It is impossible to say whether the brow ridges are prominent or not since the forehead is hardly visible in the depths of the ice.

The hairs are dark blackish-brown in colour. Their length and density have already been dealt with. Facial hairiness is very slight. But one can easily see hairs in the nares (*the nostrils*), some on the brows, and a few lashes along the eyelids.

Two further points attract the attention. The chest is particularly naked; not, certainly, to the extent seen in anthropoid apes, but nearly so – the hair, growing sideways from a median line on the sternum, leaves a highly visible skin area. The dorsum of the foot is as abundantly covered with hair as the legs are, which is not seen in anthropoid apes, who actually have almost naked feet.

Following this very detailed anatomical description, Heuvelmans next considers what the specimen might be. The first two of his five possibilities reveal that he is fully aware of the potential for a hoax. They are that it is 'an artificial, entirely manufactured object' or 'a composite, produced by assembling members and organs taken from creatures of different species'. This is a well-known skill of Chinese fakers who, for centuries, have sewn together reptile and fish parts to create 'mermaids' for the curiosity trade.

Heuvelmans dismisses the first of these – that it is an artificial wax or plastic model – by its sheer detail: 'down to the pores of the skin, and one had with infinite patience planted millions of hairs, one by one and at the appropriate angle, it is certainly more perfectly and cleverly done than any I would have thought possible.' The second hypothesis is rejected, though with less conviction, by the absence of visible sutures and (like the first) by the argument that a deliberate hoaxer would have made the model resemble far more the traditional image of a prehistoric man than was the case with this specimen.

Heuvelmans goes on to consider the possibilities that present themselves, assuming the specimen to be genuine. It is a normal man, but from a known – or unknown – abundantly hirsute ethnic group, or an abnormal man suffering from hypertrichosis (a condition where the body is covered in hair) or some other defect. These further explanations are systematically set aside and Heuvelmans is left with only the fifth possibility still standing:

that the specimen belongs to a new hominid species.

Having come to that conclusion, and absolutely convinced that they were not the victims of an elaborate hoax, Heuvelmans and Sanderson set about trying to ensure that the specimen would be preserved for detailed scientific examination. While Sanderson attempted to return to Winona with X-ray equipment to obtain a view of the skeleton, Heuvelmans set about pressuring his extensive network of high-level contacts to help him secure the body. The first of these was the anthropologist Carleton S. Coon whom both Sanderson and Heuvelmans visited at his home on Long Island, New York on 28 December 1968. After examining the photographs and scale drawings of the specimen, Coon declared himself satisfied that this was a hitherto unknown species of ultra-primitive hominid, and suggested further tests to confirm this view.

Heuvelmans and Sanderson also had to think about how and when to publish their findings, aware that Hansen was getting increasingly nervous about the consequences of exposure. Heuvelmans was quick to set the wheels in motion. By the end of January 1969 he had written a scientific report and secured its rapid acceptance for publication in the February issue of the Bulletin of the Royal Institute of Natural Sciences of Belgium. Now confident that his paper would soon appear, on 9 February 1969 he wrote to Dr John Napier, Director of the Primate Biology Program at the Smithsonian Institution in Washington DC. Unusually for the archive, which normally only contains replies to correspondence, there is a handwritten draft of this important letter.

Dear John,

I take it that you have heard from Ivan (*Sanderson*) about the most extraordinary discovery we made last December. Herewith enclosed you will find an English version of the preliminary notice I wrote about it for Bulletin de l'IRSN de Belgique, which will be published in a couple of weeks.[1]

You can imagine what the Institute's director and I had to go through, on both sides of the ocean, to be able to publish this notice that fast. But I think this was an essential thing to do.

You know how complicated the whole situation is, because of the reluctance of the owner of the specimen to have it examined properly.

At this point in the letter Heuvelmans reveals that Sanderson has been kept in the dark about the forthcoming notice in the Belgian science journal.

If I just let Ivan handle the whole business, as he demands it, this discovery will be first published in 'Argosy' (*a pulp magazine to which Sanderson was a regular contributor*) which has been associated with quite a lot of blatant hoaxes. Should this happen, nobody would take this case seriously, and I am afraid the specimen will go on decomposing (*BH and ITS had smelt a faint odour of decaying flesh during their examination of the body*) and will eventually be disposed of and for ever lost for Science.

It is for this reason that I have had a scientific notice published immediately, although it will take some time before this will reach the proper American institutions. Now I am trying to have the story publicized by 'LIFE' Magazine, before 'Argosy', to attract the attention of both scientists and authorities, and public opinion, so that the caretaker of the specimen would be forced to deliver it for a thorough scientific study. Unfortunately LIFE will not publish until a scientific examination has taken place, so we are faced with a vicious circle.

He then repeats his concerns about Sanderson's involvement.

I want to stress that Ivan and I disagree on how to handle the whole situation, as far as both publicizing the story concerning the specimen are concerned. Moreover, although Ivan has to realise that he has no scientific reputation and that he does not care, he does not realise that whenever he appears on the scene or whenever his name is mentioned, nobody takes the matter seriously anymore.

Heuvelmans ends his letter with an appeal to Napier.

We (*BH and ITS*) discussed the matter on the phone yesterday and decided that you were probably in the best position for having the right steps taken to secure the specimen, for instance with the help of Dillon Ripley (*the head of the Smithsonian*).
    With kindest regards,
      Very sincerely,
      Bernard

Heuvelmans' letter galvanised Napier into action, as his reply of 14 February makes very clear.

Dear Bernard,
    I very much appreciated your letter and the manuscript (*an English version of the forthcoming publication*). Let me say straight away, that I have already taken official action. Several days ago I contacted Sid Galler, the Secretary for Science, who is an understanding man; he is now in possession of all the facts and will approach Dillon Ripley. (I could approach Dillon Ripley direct but, administratively, it is wiser to go through Galler). The first question Galler put to me was:- 'Has anybody informed the F.B.I.?' I was able to give him a documented answer – that I.T.S. had reported the matter on Jan.18th. With this assurance, Galler felt Ripley could act – by calling Hoover (who else?) on the phone (*J. Edgar Hoover, the legendary head of the FBI*). So be reassured that the Iceman is – or will be shortly – recognised at Federal level.

Napier then confirms to Heuvelmans that the Smithsonian was prepared to carry out a thorough scientific examination of the Iceman if they could get the body to Washington. This proved to be much easier said than done.

In a statement responding to press enquiries, Napier confirms the Smithsonian's interest in examining the Iceman but concedes that: 'There is very little we can do to get hold of the specimen, if the owner does not wish us to examine it. What remains, however, is to give this matter wide publicity in the press and other media in the hope that the owner will eventually respond to public pressure and make this creature – if it really exists – available for scientific study.'

The note of institutional caution is even more evident when Napier responded to the question of whether the Smithsonian considered the specimen was genuine.

> The Smithsonian is prepared to keep an open mind. However, as we have already pointed out to the Press on the basis of information received, the creature may well be only a model. Nevertheless it must be conceded that a reputable scientist, Dr Bernard Heuvelmans and an experienced taxidermist, Mr Ivan. T. Sanderson, have both expressed their views that the creature is real. It is not possible at the present time to offer any firm opinion.

That Napier, true to his word, did persuade Dillon Ripley to act is shown in his letter to Hansen of 13 March.

> Dear Mr Hansen,
>
> One of our scientists has received a reprint of an article published recently in a scientific journal, The Bulletin of the Royal Institute of Natural Science of Belgium. The article written by Dr Bernard Heuvelmans describes a specimen preserved in ice that is at present in your possession.
>
> From the description and the photographs it would seem that this creature is of great interest to the scientific community. I am writing to ask if you would be willing to give the Smithsonian's scientists the opportunity of examining the specimen so that they may make their own judgement on its significance. As you may know this Institution has a tradition of studies in physical anthropology extending back over one hundred years.
>
> I should greatly appreciate your cooperation in this matter

which may prove to be an outstanding contribution to human knowledge.

> Sincerely yours
> S. Dillon Ripley
> Secretary

I have my doubts that this polite request from an unknown administrator from some far-off museum would suddenly persuade Hansen, the Midwest showman, to change course and allow the body to be removed to Washington. And so it proved. His reply to Ripley is dated 20 March 1969.

> Dear Mr Ripley,
>
> The specimen mentioned in Dr Heuvelmans' report is presently in the hands of its owner who insists, for various reasons, to remain anonymous.
>
> Due to this report which was made without the owner's knowledge or permission, I believe it would be very difficult to persuade him to amend his present position and permit any form of scientific evaluation.

At this point in his letter Hansen introduces a new dimension to the case when he writes:

> We have framed an illusion show for 1969 which, in many respects, resembles the specimen photographed by Dr Heuvelmans while a guest at our ranch. I cannot say with any certainty if the original exhibit will ever be presented again as a public attraction.
>
> Should there be any change in the owner's present position I will be happy to advise.
>
> Thank you for your interest.
> Sincerely yours,
> MIDWEST EXHIBITS
> F.D. Hansen

Hansen had changed his story. This time he announces that he was about to go on tour with a dummy version of the original Iceman, billing it as a deliberate hoax. And indeed he did put something on show in a shopping centre in St Paul, Minnesota in May 1969 and later in Grand Rapids, Michigan. It was almost certainly not the same corpse that Heuvelmans and Sanderson had examined back at the Hansen ranch in December of the previous year. Photographs of the new exhibit taken by *Time-Life* showed many differences between this replacement and the original.

In the interim, the increasingly suspicious Napier had been making enquiries in California and had located a wax museum whose owner had allegedly been commissioned to apply hairs to a latex model made for Hansen in Hollywood two years before, in 1967. That discovery seems to have swung the balance for the Smithsonian and they decided to wash their hands of the whole business and publicise their withdrawal in a press release. Napier explained their reasons in a letter to Heuvelmans, a letter that must have been very difficult for him to write. The key sentence is this one:

> I felt it was time to get the Smithsonian off the hook and we gave out a press release saying that our own information leads us to the conclusion that the Iceman was a carnival model and not the real thing.

In a second letter to Heuvelmans, Napier adds that the decision to announce the Smithsonian's withdrawal from the investigation was partly due to the fact that they had experienced a run of bad publicity in connection with other projects and could not risk a third embarrassment. The risk of looking foolish also seems to have been behind the FBI decision not to take any further action. Once there was even the faintest whiff that the Iceman might be a hoax, J. Edgar Hoover was certainly not going to risk being made a laughing stock by raiding the Hansen ranch and recovering a latex model.

Not surprisingly, Heuvelmans was furious, as his letter to Napier shows. As a general rule he did not keep copies of his own letters, but this is one of the occasions when he did. I have had to rely on his heavily edited handwritten draft full of crossings out. I was unable to make sense of everything, but here is what I deciphered:

Dear John,

When I came home last Sunday, I found your letter posted on 12th May which really threw me down. To receive the 'brush-off' before any serious investigation, I must say it is difficult for me to remain civil. Saying that it was done 'to get the Smithsonian off the hook' seems to me a very euphemistic way of putting the fact that this Institution is deserting the battle just because the enemy slightly frowned.

But, please John, try to put yourself in my place. I have examined the specimen very carefully for 11 hours (eleven!) over three days and am absolutely positive about its being genuine. In my scientific notice I had to consider the possibility of a fake because it was theoretically one of the possibilities, but practically I can assure you that I cannot have been fooled. Just imagine that you would have scrutinised a specimen for such a long time, even through an inch-thick layer of ice – crystal clear at some places – and you would have concluded that it was genuine. How would you feel if somebody - if most people – told you that you are mistaken. This was not just a photograph, or a film, this was a 3 dimensional thing that I have been able to study very thoroughly, under many different angles, with the aid of a floodlight.

In the meantime, Napier had composed a detailed report on the affair for his boss at the Smithsonian, Dillon Ripley, entitled 'Windup on the Iceman' that gives his reasons for believing the Iceman to be a fake. Among these is the admittedly curious reluctance of Hansen to let the corpse be properly studied.

If it is real, why hasn't Hansen stated it in so many words. He has everything to gain by it being a real specimen. If the Iceman is real, Hansen (or the mysterious owner) has in his possession the most unusual natural history specimen in the world. If it is money he wants then he need never be poor again. If he is not interested in money then, by presenting the specimen to science, he would become enshrined forever in the halls of the great discoverers; and what is more he could <u>still</u> take the model round the carnivals and be sure of a living as 'the man who discovered the missing link.' It is completely illogical to me that a showman should take a real specimen round the fairgrounds at 35 cents a peek, when by making it available to science he would make a million dollars, become immortal – or both.

Was the Minnesota Iceman a fake or was it genuine? On the one hand we have Heuvelmans' insistence that he cannot have been fooled and, on the other, the perplexing refusal of Hansen to hand over the body for proper scientific investigation. Unsurprisingly, the archive contains abundant 'conspiracy theories' – for example that the US government seized the corpse so as to avoid having to respond to a request from Russia to repatriate the body of one of its citizens illegally seized by the Americans (remember this was all taking place at the height of the Cold War).

In any event, Heuvelmans wrote to Napier and to Dillon Ripley trying to get the case re-opened, but to no avail.

Nothing was heard from Ivan Sanderson until a letter of 6 June 1969 in which he reminds Heuvelmans:

As I told you, Bernard, in one of my telephone conversations with you . . . I only wish to God that you had not published before all the investigations . . . had been given time to be prosecuted. Hansen knew about your article almost on the day that it was published in Belgium and that was when he disappeared for three weeks with the specimen. He phoned me and admitted frankly that, as a result, it had been hidden for all time, and that they were having extensive changes made to the wax model.

I don't think we will ever get the original now but naturally I still have hopes to vindicate your brilliant analysis. For my part, I am publishing also, but merely as a reporter and an amateur because, after all, I only have M.A.s (*Master of Arts, from Cambridge, England but in fact in the scientific disciplines of geology and ethnology*). I shall keep you informed in the hope that you will be willing to criticise as always. As I told you in the car coming back through the snowstorms from Minnesota, it could only be FATE that brought you here when all this extraordinary business happened.

Yours as ever

Ivan

But relations between Heuvelmans and Sanderson were never repaired. Heuvelmans saw Sanderson's reputation for sensation as an obstacle to the scientific respectability that he craved, and Sanderson blamed Heuvelmans' publication for Hansen's disposal of the original Iceman. Sanderson continued to blame Heuvelmans for the rest of his life, as this extract from a letter written in January 1972 to Tom Hall of *The Sunday Times* makes all too clear:

One little point: it was Dr Bernard Heuvelmans who went ahead against his promise to me – let alone Hansen – that he would not publish, and I was even holding up my purely popular piece for ARGOSY having given said promise. I was keeping the famous 'open mind' until I could get the damned thing out of the ice, and came damned near getting permission from Hansen (if not this mysterious owner Mr X., if he exists), when that bloody fool Heuvelmans bulldozed the Belgian Academy to rush his version into print. I may add that he never even told me he had written it, though he had been a guest in my house up to the time he (I presume) started writing it. Well, he got his just desserts, as he lost his press card on the one hand and has been

turfed out of the scientific fraternity on the other. That's taught
him the validity of 'off the record' disclosure.

After that, the only letters in the Sanderson file, once bristling
with their frequent correspondence, are from Ivan's daughter
informing Heuvelmans of her father's death in 1973.

Mainstream scientific involvement in cryptozoology ended with
the collapse of the Minnesota Iceman affair. No professional
scientist would go near the subject for fear of ridicule and it slid
into the abyss where it resides to this day. Heuvelmans was never
the same again either. He never retracted his absolute conviction
that the body he examined had been genuine, and he never
forgave the 'scientific establishment' for disbelieving him. From
then on he abandoned all ambition to be accepted by the academic
circle that had rejected him and immersed himself in his celebrity
role as the 'Father of Cryptozoology', a guru surrounded by
admiring and gullible acolytes. He died, after a long illness, in
Paris on 14 January 2001, cared for by his former wife, Alika
Lindbergh, whose unconventional relationship with Heuvelmans
endured until the very end.

There is a postscript to the story of the Minnesota Iceman. In
July 2013 a latex model, advertised as the one that had fooled
Heuvelmans and Sanderson, resurfaced for sale on eBay with a
reserve of twenty thousand dollars. Though the vendor's identity
was not revealed it was reportedly purchased by Steve Busti for
display in his 'Museum of the Weird' in Austin, Texas. It was
due to be exhibited on loan to Loren Coleman's International
Museum of Cryptozoology in Portland, Maine, but Loren tells
me that, as of June 2014, it has yet to materialise.

9

Clutching at Straws

In the years before the Minnesota Iceman fiasco, when Bernard Heuvelmans was actively expanding his collection of press cuttings and correspondences, he was collecting evidence for the existence of 'animals unknown to science'. Though he was desperately disappointed by the reaction of the scientific establishment to what he believed was the body of a new hominid or a genuine Neanderthal survivor, he took immense encouragement from the documented and widely accepted discoveries of new species that had been made in the nineteenth and twentieth centuries, and which still continue in the twenty-first.

A coelacanth, the lobe-finned fish thought to have become extinct over sixty million years ago, was caught in 1938 off the Chalumna estuary in the Eastern Cape province of South Africa. It was discovered in the catch of a local fisherman by Marjorie Courtney-Latimer, the curator of the museum, although it was a local chemistry professor James B.L. Smith who was the first to realise its importance as a modern survivor from an ancient genus. It was given the species name *Latimeria chalumnae* to recognise its discoverer and the River Chalumna where it was caught. Sixty years later, in 1998, a second species of coelacanth was discovered from the Indonesian island of Sulawesi after a tourist couple had managed to photograph a specimen on sale at a local market only moments before it was bought for someone's supper.

The okapi, a secretive and beautifully marked forest-dwelling relative of the giraffe, was only a myth in the West until parts of a carcass were sent to London in 1901 by Sir Harry Johnson, British Governor of Uganda, from what is now the Democratic Republic of the Congo. Henry Morton Stanley's expedition to Central Africa in the 1870s heard about a mysterious animal with striped flanks living in dense forest, but he never saw one. There are estimated to be about twenty thousand okapis presently living wild in the forests of Central Africa. Appropriately, the okapi was adopted as the emblem of the International Society of Cryptozoology, of whom Heuvelmans was the founding president.

After being rescued, a sailor who was stranded on the Indonesian island of Komodo for several months told stories of a dragon that ate pigs, goats and even attacked horses. Of course, no one believed him, at least not until they were confirmed by a Dutch colonial administrator in 1910. An expedition in 1927 brought out two live specimens of 'Komodo Dragons' as they become known, one of which was the star attraction at the opening of the new Reptile House at London Zoo the following year.

Even an animal the size of a small buffalo remained 'unknown to science' until 1992 when the saola (*Pseudoryx nghetinhensis*), a forest antelope weighing up to 100kg, was defined from three sets of horns found in the huts of hunters in the jungles of Vietnam. Despite an intense search for a live specimen, it took more than twenty years before the first saola was photographed in the wild, by a camera trap in September 2013. Live saola have occasionally been captured by villagers but all have quickly died in captivity.

No one has seriously suggested that the yeti and Bigfoot are large hooved animals, but the following serendipitous discovery did provide a popular candidate for these mythical creatures. In 1935 the German paleoanthropologist Ralph von Koenigswald, based at the time in Java, was in an apothecary's shop in Hong Kong looking at the 'dragon's teeth' for sale as an aphrodisiac. It was not that von Koenigswald was in the market for performance-enhancing drugs. He was there because he knew that interesting fossils from China's abundant limestone caves sometimes made their way into such displays. Casually he picked out a molar tooth that caught his eye and, after purchasing and studying it more closely, concluded that it had probably belonged to a giant ape. Further finds in China, including jawbones, supported this and a new species was confirmed. Von Koenigswald named it *Gigantopithecus blackii* – the first part meaning 'giant ape' and the second in honour of his late colleague and friend Davidson Black. The assumption was that *Gigantopithecus* had died out between two hundred thousand to a million years ago. It is difficult to imagine a tooth that old making a palatable aphrodisiac. In my experience of drilling into fossil teeth to extract DNA, after a thousand years in the ground, the organic parts have been completely replaced by stone. It would have been a gritty mixture to convert into a medicine, even in an emergency.

If it survives, as some cryptozoologists believe it might, *Gigantopithecus* would certainly be big enough to match up with some of the more flamboyant descriptions of yeti and Bigfoot, with estimates putting an adult male standing at ten feet tall and weighing 1,000 to 1,200 pounds. *Gigantopithecus* is certainly the favourite among many yeti and Bigfoot hunters. The other primate in the frame is the orang-utan and, of course, our own hominid cousins. Of the three known great apes, only the orang-utan is found in Asia. The last surviving refuges of this arboreal ape are the forests of Sumatra and Borneo, but in the past the animal was far more widespread with a range extending, according to fossil evidence, through Malaysia and into China, where they were much larger than their modern extant counterparts farther south. John Napier, in his systematic review of the possibilities, considered that the orang-utan could well survive in the high forests of the Arun Valley and puts it at the top of his shortlist of yeti candidates. And then there are the hypothetical unknown primates whose characteristics are, by definition, unknown. But even a hitherto unknown species, as we shall see, cannot escape recognition by DNA.

While most of the evidence for the existence of yeti and Bigfoot has come in the form of eyewitness statements, and the casting and analysis of eponymous tracks, there have been some half-hearted attempts in the past to retrieve information from organic materials recovered from locations where these creatures are thought to have left them. There have also been investigations into relics removed, with or without permission, from Himalayan monasteries.

While some professional scientists, like John Napier and in more recent times Jeff Meldrum, absorbed these enquiries into their own research programmes, mainstream scientific involvement was more common in the days of the great expeditions, and before the Minnesota Iceman fiasco. Tom Slick, for example, assembled a panel of consultants to examine material brought back from his Himalayan expeditions. For the 1958 yeti-hunt he recruited a panel of twenty-one consultants including the eminent, if controversial, anthropologist Carleton S. Coon from Penn State University and his colleague Paul Baker, primatologist William Osman Hill from the London Zoological Society, hair expert George Agogino from the University of Wyoming along with Bernard Heuvelmans and Boris Porchnev. When consultant panels of such prowess are created it is often to add prestige to a prospectus, usually with the goal of raising financial support. Very few actually do any real work. This seems to have been the case with many of the Slick consultants, but some did at least glance at the materials that were brought back from the Himalayas. At that time forensic techniques for examining such organic material were in their infancy, meaning that conclusions were vague and often no more than personal opinions that were liable to differ between experts.

Take, as an example, the different conclusions reached about some yeti droppings brought back by Peter Byrne from the 1958 Arun Valley expedition. George Agogino, who was responsible for distributing samples of the droppings among the consultants, sent small portions to William Osman Hill at London Zoo and to Ralph Izzard, who led the 1954 *Daily Mail* yeti expedition. Izzard was primarily a journalist, so quickly replied that he did not have the skills to offer an opinion. Osman Hill looked at the droppings under a microscope and came to the conclusion that they came from a herbivore. Agogino, on the other hand, thought these were the droppings of a carnivore, perhaps a wolf.

Franklin G. Wallace from the University of Minnesota, who was a first-rate parasitologist, wrote that the droppings were 'not human; most improbably primate; most probably from a sheep or other herbivore'.[1] Wallace based his conclusion on three intestinal parasite eggs that he found in the faeces. Parasites are so well adapted to their hosts that they are usually completely restricted to one species of host. That doesn't mean, for instance, that we humans will never be bitten by a cat flea, but a cat flea will not survive for long in the face of competition from human fleas that are far better adapted to living on us and sucking our blood than their feline counterparts. To an expert like Wallace, the eggs of different parasites can be differentiated and the host species identified just by the eggs of the parasites found in the faeces. Wallace thought that the three eggs he located in the 'yeti' sample were not from a human parasite species but from a sheep – an opinion that was especially firm in the case of one particular egg.

Bernard Heuvelmans disagreed with Wallace and cited a separate examination of the droppings by the prolific Antwerp parasitologist, Dr Alex Fain, who, according to Heuvelmans, found the egg of a species in the genus *Trichuris* which he was unable to identify. Heuvelmans made much of this, inferring that a new species of parasite must therefore mean that the droppings belonged to a new and as yet unidentified host species. Heuvelmans' remarks have all the hallmarks of a man clutching at straws. It is also a good example of the irrational and irritating claim, often repeated by cryptozoologists, that if a sample is not positively identified it means it must have come from a new, unidentified species, *ergo* a yeti or Bigfoot. As far as I am aware, prolific author though he certainly was, Dr Fain did not publish his yeti findings.

Finally in 1979, the same faeces were re-examined by Dr Anne Porter, an associate of Osman Hill. She identified mammalian hairs, probably from a small rodent, parts of a caterpillar and a grasshopper, from which she concluded that the donor of the droppings probably ate frogs. Again, as far as I can find, this was never published and the only reason we know about it at all is a postcard she sent to Tom Slick.

The ways in which the yeti droppings were examined and the results reported are very familiar. This is not the careful work of scientists committed to finding an answer, but has all the signs of someone having a quick look at a sample with dubious provenance. No publications resulted and the results themselves, or rather opinions, varied from sheep, to wolf, to an animal with a penchant for frogs, to the best they could hope for, an 'unknown' species. As usual, all contrary evidence was ignored and the 'unknown' species hinted at by Dr Fain's interpretation of a parasite egg was triumphantly presented to the world as proof that the yeti was alive and well and defecating somewhere in the Himalayas. It strikes me as ironic in the extreme that this proof, as Heuvelmans claimed it to be, of a gigantic primate roaming free in the wilderness rested on a single parasite egg smaller than a pinhead.

On the positive side, animal droppings have been used successfully as a source of DNA, for example in studying the movement of grizzly bears in North America on both sides of the Rockies. I used droppings myself back in the 1990s to see if the unlikely story that all golden hamsters are descended from a single female was true. It was. However, there are drawbacks when it comes to trying to identify the depositor of the droppings. The bear and hamster work used a genetic protocol based on knowing details of their genomes. The technique was tweaked to selectively target bear or hamster DNA and ignore the rest. So it didn't matter what the bear was eating as only bear DNA from cells lining the intestines that had been sloughed off and appeared in the droppings was targeted. So the DNA from grubs, berries, rodents or even elk in the bear's diet is never picked up by the reaction. But if you don't know the animal, then you can't design the reaction to suit the sample, and therefore risk picking up DNA from the food eaten and the resident parasites, rather than the animal itself. Applied to the same droppings that were tested from the Slick expeditions, there's a risk the yeti would be identified as a frog or even a giant grasshopper.

This problem disappears when hair is used as the source material instead of droppings. Only mammals have hair, so that's a good start. We will go on to see why hair was the first choice for my genetic analysis, but hair also has several anatomical features that can be used to identify the species of the owner. Hair is made in the root, or follicle, and extruded through pores in the skin. Each strand is roughly cylindrical in most animals and comprises three concentric layers, which can easily be distinguished under a light microscope. The central core of a hair shaft is the medulla, surrounded in the next layer by the cortex, and finally, on the outside, by the cuticle. Each layer has its own characteristics that vary between species, or at least between different families of mammal. Sometimes, for example, the hair may have a very small or even non-existent medulla, in which case the cortex will make up the major part of the hair's overall thickness. In other species the medulla takes up most of the space. The appearance of these two components also varies between species. It is the cortex that contains the melanin pigment granules and their appearance – rough, circular, streaky and so on – are additional signs that help identify the species. The medulla too can have a characteristic appearance in different animals. For instance, deer are very easy to identify from the appearance of their very cellular medullas, which look as if they are full of air bubbles. The outermost layer, the cuticle, is made up of overlapping cell remnants that resemble scales. The pattern of the cuticle scales is also part of the diagnostic analysis. Although the general scale pattern can be seen under an ordinary light microscope, it is hard to catch the detail without using a scanning electron microscope with much higher magnification and only after the cuticle has been coated with gold particles to produce the exquisite contrast.

However, the one major drawback of these techniques is that getting a firm species identification just by the appearance of a hair is extremely difficult and requires a very high level of skill and experience in the microscopist. Another complication is that hair differs a lot depending on where on the body it's from. I realised this when I spent time with the US Fish and Wildlife hair morphology experts Bonnie Yates and Cookie Smith at their forensic laboratory in Ashland, Oregon. Both at the top of their profession, they told me that they would never give a firm species identification on the basis of a single hair. Bonnie explained that many animals have at least two different types of hair. In bears and deer, for example, one in a hundred or so hairs is much longer and thicker than the rest. These are the guard hairs that protrude beyond the main pelt and are there to protect the finer and more densely packed underhairs from damage. When they are moving through the forest at night these animals can feel when the guard hairs touch an obstacle like a tree trunk and can move aside to avoid it.

Though primates do not have guard hairs, the hairs vary a lot depending where they are growing on the body. In men, for example, beard hair is much thicker than head hair. Pubic and underarm hairs are different again. And, of course, there is a huge variation between individuals, which may or may not reflect their ethnic origin. On the whole African hair has a flatter cross section and the result is a tight curl. Asian hair, on the other hand, has an almost perfectly round cross section and is very straight. European hair is somewhere in between, but with enormous variation between individuals.

With all these provisos and complexities, it is no surprise that identifying a new species, like a yeti, from the appearance of a hair sample is a tough call. In fact, without a 'type specimen' to compare, it is impossible. All that can realistically be expected is an indicative diagnosis such as 'the hair has primate charac-teristics' or 'it is from an unknown species'. The last conclusion is the most fallacious, especially when it is a distortion of a professional judgement that, on the available evidence (often just a single hair), a positive identification simply cannot be made. There is a danger that this cautious professional conclusion under-goes the slippery transition to become the hair of 'an unidentified animal'. Bonnie Yates, for one, has had quite enough of this sleight of hand and no longer accepts samples for identification from cryptozoologists.

Dr Henner Fahrenbach, on the other hand is more sympathetic. A biomedical scientist originally from Oregon, now living in Arizona, he has studied sasquatch hairs for many years and has compiled a list of features to aid their identification by microscopy, though understandably without the benefit of a type specimen. These are his thoughts, written for the International Society of Cryptozoology:

Generally, sasquatch hair has the same diameter range as human hair and averages 2 to 3 inches (5–8cm) in length, with the longest collected being 15 inches (38.1cm). The end is rounded or split, often with embedded dirt. A cut end would indicate human origin.

Sasquatch hair is distinguished by an absence of a medulla, the central cellular canal. At best, a few short regions of a frag-mentary medulla of amorphous composition are found near the base of the hair. Some human hairs also lack a medulla, but the current collection of 20 independent samples with congruent morphology effectively rules out substitution of human hair.

The cross-sectional shape and color of sasquatch hair is uniform from one end to the other, in keeping with the char-acteristics of primate hair in general. There are no guard hairs or woolly undercoat and the hair cannot be expected to molt with the seasons. Hence, hair collections are invariably sparse in number.

Despite a wide variety of observed hair colors in sasquatch, under the microscope they invariably have fine melanin pigmentation and a reddish cast to the cortex, presumably a function of the pigment pheomelanin.

I contacted Dr Fahrenbach to ask if I could carry out a DNA analysis on those hairs in his collection that had satisfied his own criteria and which, in his opinion, were most likely to have come from a sasquatch. As we will see later, he was happy to oblige.

The few hairs from the Slick expeditions that were examined were either immediately dismissed as irrelevant, in that they clearly didn't come from primates, or fell into the category of 'perhaps primate'. Very unfortunately, neither the yeti-hunters who collected the hairs nor the scientists who examined them realised the potential for unambiguous species identification beyond the 'perhaps primate' limit. No one thought about DNA then. The hairs were not treasured as the key witnesses they have now become, and they are mostly lost. I have tracked down a few, but not as many as I would have liked.

No such 'perhaps primate' category is possible with a well-conducted DNA analysis. This point was clearly not appreciated by either the speaker or the audience of cryptozoologists at 'Weird Weekend' when I listened to the astonishing adventures of Adam Davies and his companions in search of the elusive *orang-pendek* in the jungles of Sumatra. On his latest expedition, Adam had found a short hair close to a site where one of his companions had spotted the diminutive primate. He included in his talk the results of a DNA analysis which he told the audience had shown characteristics of both human and non-human primate DNA, putting it straight into the 'perhaps primate' category, just like the hair morphology studies of yesteryear. I had to restrain myself when it came to question time, as this was patently impossible. He was reporting a study using mitochondrial DNA which gives a precise sequence that can be unambiguously assigned to an extant species, or, as we will discover, an unknown one. It cannot share the characteristics of two species. 'Weird Weekend' was not the right forum to have this out in the open, but it did show me that cryptozoologists were not sufficiently versed in genetic analysis to realise that what was being said about the DNA result was quite impossible.

10

Our Human Ancestors

The two most promising candidates as the source of our anomalous primate, be it yeti or Bigfoot, are the giant ape *Gigantopithecus* and other human species, like Neanderthals, favoured by Heuvelmans and Porchnev. There are several books that debate these two themes, often in great detail and with great passion, but which tend to forget that they have no convincing evidence for either. Both species are generally held to be extinct, though as we have seen the hope among cryptozoologists is that there are surviving pockets in remote parts of the world. Until very recently, there was no serious intellectual platform to support either theory. While nothing has happened to enhance the prospects of *Gigantopithecus* as the biological incarnation of yeti or Bigfoot, there has been spectacular progress in discovering the complexity of our own human evolution. It now seems as though our *Homo sapiens* ancestors shared the planet with several other human species and even interbred with them. The notion that there could be parts of the earth where these other humans survive to this day, either as a completely separate species or as a type of genetic hybrid, does not seem anywhere near as ridiculous as it once did.

The currents of thought in the matter of human evolution have flowed back and forth ever since Darwin convinced the world (or most of it) that we evolved from other species over a very long period rather than being created in our present form. The discovery which began our appreciation that we have not always been the only human species on the planet was made in 1856 when a very odd skull was found in a limestone cave near Düsseldorf, Germany. Among other unusual features, it had a receding forehead and very prominent brow ridges. At first it was dismissed as a freak, a mutant with some sort of deformity. However, when other very similar skulls began to turn up from excavations, first in Gibraltar, France and Belgium and later in the Middle East, it slowly dawned on antiquarians, the nineteenth-century predecessors of today's palaeontologists, that these were not diseased skulls at all, but belonged to another type of human, eventually called Neanderthal after the Neander Valley (*thal* in German) where the first skull, the type specimen as it is called, was discovered. There followed over a century of often acrimonious debate as to whether Neanderthals were our own ancestors or whether they belonged to a different species now extinct. Judging by the age and distribution of Neanderthal and *Homo sapiens* fossils, the two species did share the same geographical range in Europe as well as further East. The long debate was finally settled in the 1980s when both genetic and careful anatomical analyses concluded that Neanderthals were a completely separate species and that our own ancestors, 'newly' arrived in Europe from Africa 40,000 to 50,000 years ago, had replaced them.

The argument appeared to be settled after the successful recovery of DNA from the Neanderthal type specimen in 1997 which showed that its mitochondrial DNA sequence was quite unlike that of any modern human. The intellectual dominance of the 'Recent Out of Africa' camp lasted until 2010, when the same German team that had published the first Neanderthal mitochondrial DNA in 1997 declared, to an astonished and admiring world, that they had succeeded in sequencing the entire genome (more or less) from a particularly well-preserved set of Neanderthal bone fragments from Vindija cave in Croatia.[1] The 2010 paper did not entirely reverse the earlier conclusion that we are the descendants of a separate species rather than merely modified Neanderthals but it did add an intriguing twist. After comparing the Neanderthal DNA sequence with modern *Homo sapiens* genomes from Europe and Africa, the authors concluded that Europeans, though not Africans, were partly descended from Neanderthals rather than being genetically entirely separate. All Europeans tested, then and since, have inherited between 2% and 4% of their nuclear DNA from a Neanderthal ancestor. The explanation given was that this could only have occurred through interbreeding. This shocking news, for it was a shock to all concerned, has taken time to digest. Even the principal investigator, Svante Pääbo, didn't believe it at first, instructing his colleagues to double-check the data and the calculations.

The reasoning is complex and depends on three-way comparisons of the sequences from Neanderthal, European, African and also Asian genomes. This is not something one can do on the back of an envelope and it required the pooled expertise of many of the world's top brains in the new field of bioinformatics, assisted by a cluster of 256 computers, to score and compare the sequences of the billions of DNA fragments streaming from the new generation of DNA sequencing machines. The basic question asked of each of the Neanderthal DNA fragments was this: is it more similar to the same fragment in a modern European, or a modern African genome? The null hypothesis was that the Neanderthal DNA fragments would most closely resemble an African *Homo sapiens* sequence 50% of the time and a European *Homo sapiens* sequence 50% of the time. This would indicate that there was no interbreeding. This hypothesis is based on the assumption that Neanderthals shared a common ancestor with all *Homo sapiens*, so any changes in the DNA sequences between humans and Neanderthals must have occurred since that split from the common ancestor. As the time elapsed since the split from the common ancestor is the same for all modern humans, whether Africans or Europeans, and changes accumulate at a steady rate, the null hypothesis predicts that the matches scored in Neanderthal v. African and Neanderthal v. European sequence comparisons would be equal.

But they are not. The Neanderthal fragments were slightly more similar to their equivalent fragments in European than in African genomes. The null hypothesis predictions of 50/50 balance were shifted faintly in favour of a closer genetic match to Europeans than to Africans. In one comparison, for example, Neanderthal fragments had closer matches to a modern French genome 52.5% of the time, and to a Yoruba from Nigeria in 47.5% of comparisons. Though the difference is small, so many millions of comparisons were made for each analysis that the final result is highly statistically significant.

Less clear, though, are the alternative interpretations. For example, if Neanderthals were descended from an ancient African population in, say, East Africa that was also the source of later *Homo sapiens* dispersals, that shared ancestry may have contributed to the closer genetic affinities to Europeans apparent in the sequence comparisons. Or could it be that the genome of *Homo sapiens* changed to meet the much colder conditions of Ice-Age Europe in the same way as the Neanderthal genome became adapted hundreds of thousands of years earlier? An intriguing paper published in 2011 noted that the DNA segment with one of the strongest similarities between Neanderthal and modern Europeans is located in the part of the genome that controls the immune response, vital for fighting infections.[2] Having the same immune response genes as Neanderthals, perhaps through interbreeding or maybe through selection of pre-existing genetic variants, could well have been critical for the survival of *Homo sapiens* in Europe in the face of resident Neanderthal pathogens.

Svante Pääbo, who led the Neanderthal Genome Project, has written a candid first-hand account which covers all these possibilities, and also reveals the excruciating complexity of the venture.[3] It wasn't only the bioinformatics that was difficult. Getting reliable data from the tiny percentage of Neanderthal DNA remaining in the Vindija bone fragments in the face of overwhelming amounts of bacterial and modern human contamination was a gruelling business indeed, taking its toll on budgets, collaborations and even on Pääbo's health.

One immediate puzzle that needed explaining was this. How is it that not a single Neanderthal mitochondrial DNA has been ever been found in a modern European? Several hundreds of thousands, maybe even a million, Europeans have had their mitochondrial DNA tested whether in research projects or as customers of genetic genealogy companies. Yet there has never been a whisper of Neanderthal mitochondrial DNA in any of them. Although mitochondrial DNA is inherited exclusively down the maternal line, all things being equal, we should expect the same proportion of mitochondrial DNA with a Neanderthal origin in modern Europeans and Asians as there is Neanderthal nuclear DNA. With the average amount of Neanderthal nuclear DNA in Europeans estimated to be 2.5%, then if a million Europeans have had their mitochondrial DNA analysed, which is a reasonable estimate after two decades of widespread testing, 25,000 of them would have had shown a Neanderthal result. There is not a single one.

This anomaly isn't quite as hard to explain as it first seems, though it remains something of a puzzle to me. The solution offered by statisticians is centred on the fact that mitochondrial DNA is far more likely than its nuclear counterpart to be lost as it travels through the generations. For example, a woman will pass her nuclear DNA to both her sons and daughters. However, thanks to its strict matrilineal inheritance pattern, only her daughters will pass on her mitochondrial DNA to the next generations. Although her nuclear DNA will live on in her sons' children, her mitochondrial DNA dies with them. The upshot of all this is that, on average, four times less mitochondrial DNA is passed on to the next generation compared to nuclear DNA. This same equation applies at each generation, so very soon there are fewer and fewer different mitochondrial DNAs in circulation.

Mitochondrial DNA can never be lost entirely because it is vital for aerobic metabolism. Nevertheless, after a few hundred generations it is theoretically possible for Neanderthal nuclear DNA to have got through to the present day and for all the Neanderthal mitochondrial DNA to have been lost. This eradication was never certain to happen, and the other scenario might have been that Neanderthal mitochondrial DNA did get through and lots of people had it. In *The Seven Daughters of Eve* I explain how I found that over 95% of native Europeans are matrilinear descendants of only seven ancestral clan mothers. If Neanderthal mitochondrial DNA had survived as well as its nuclear equivalent, at least one of these seven women might have been a Neanderthal. But that did not happen and, as the authors of the 2010 paper argued, it is just a matter of chance that no one these days carries the mitochondrial DNA of a Neanderthal.

The other solution was that the interbreeding which led to the Neanderthal DNA getting into the human genome in the first place was all between Neanderthal men and *Homo sapiens* women. In that frankly unlikely scenario, there would be no Neanderthal mitochondrial DNA in any of the offspring. This was diminished as an explanation when further work showed that there were no Neanderthal Y-chromosomes, which would have come from males, in modern Europeans either.

A few weeks before the Neanderthal genome paper was published in 2010, another astonishing genetic revelation found its way into the journal *Nature*.[4] Here the same team that had sequenced the Neanderthal genome announced that they had identified a new human species. Mitochondrial DNA was extracted from a fragment of a little finger bone found alongside other human remains in Denisova Cave in the Altai Mountains of southern Siberia during an excavation in 2008. The sequence showed that this bone fragment belonged not to a Neanderthal, nor to a *Homo sapiens* but to an as yet unknown human species. Following the tradition of naming a species after the place where it was found, the authors were tempted to name this as *Homo altaiensis*. They wisely recanted as all that remained on which to base a description of the new species was the fragment of finger bone, two molar teeth found nearby and the DNA sequence. There are strict rules about naming new species, as Heuvelmans discovered to his cost when he tried to register the Minnesota Iceman as *Homo pongoides* (literally *man-ape*). Heuvelmans' attempt to file his new species name without a type specimen so enraged traditional taxonomists that they lobbied, successfully, to have it struck off the official register of species.

Pääbo wisely avoided any such controversy, so there is no *Homo altaiensis*, at least not yet. The new species is for the moment known simply as Denisovan.

Another great surprise was that the Denisovan bone was in such good condition. It was no bigger than two grains of rice, but contained more intrinsic DNA than all the Vindija Neanderthal fragments put together. Exactly why this should be is still a mystery. The sample is too small to be carbon-dated so we do not know how old it is. One possibility is that it is very much younger than the less well-preserved Neanderthal fossils. Pääbo even suggests, though not very seriously, that it might be from a modern *alma*.[5] Now that would be something. It would also be vindication of a sort for Russian hominologists, from Porchnev onward, who always believed in the survival of Neanderthals. I am sure they would settle for Denisovans instead.

When the Denisovan sequence was compared to the same region in modern humans and in the six complete Neanderthal mitochondrial DNA sequences known by that time, there were twice as many differences between the Denisovan and *Homo sapiens* as there were between ourselves and Neanderthals. By this reckoning, the Denisovans were our considerably more distant relatives than the Neanderthals. An estimate can be made, based on the differences between the DNA sequences, of how long has passed since two species last shared a common ancestor. There are many provisos in such estimates and nobody relies too much on the precision of the figures but, roughly speaking, the last common ancestor we shared with the Denisovans lived about a million years ago while the same calculations split *Homo sapiens* from Neanderthals about half a million years back.

Thanks to the exceptional preservation of the Denisovan bone fragment it did not take long to get a good genome sequence, and a much better one in terms of quality than the Neanderthal. However, there were more surprises in store. A comparison of the Neanderthal and Denisovan nuclear DNA showed that they were much more closely related than the mitochondrial DNA had suggested. One possible explanation for the discrepancy is that the mitochondrial DNA in the Denisovans was actually from yet another, earlier human species with whom their ancestors had interbred. Its survival through interbreeding in Denisovans was just as much a matter of chance as the apparent extinction of Neanderthal mitochondrial DNA in modern humans, where we have the reverse outcome. The matrilineal lineage of the other ancestral species survived in Denisovans whereas most or all of the nuclear DNA had come from the other hybridising species.

There were yet more surprises to come when the signs of interbreeding between ourselves and Neanderthals, at least in Europe, was also detected between Denisovans and ourselves. Denisovan nuclear DNA was found not in natives of Europe but of Papua New Guinea, and subsequently in native Australians and Pacific Islanders, and at a slightly higher level, up to 4.8%. With some Neanderthal thrown as well, the total percentage of non-sapiens DNA in modern Papuan genomes rises to the substantial total of 7.4%. To explain the link between Denisovans living forty thousand years ago in Siberia and present-day occupants of Melanesian islands like Papua New Guinea, the only logical interpretation is that the ancestors of both human species had interacted elsewhere, probably as the ancestors of today's Papuans were *en route* to the islands of Melanesia. It is a remarkable story, completely unexpected from classical palaeontology and the sure sign, if one were needed, of the real contribution ancient DNA is now making to the understanding of our own evolution.

It is quite likely that there are more collateral hominids yet to be discovered. Interbreeding between different human species, once thought unlikely or impossible, is now all the rage, with DNA signals of mixing between the ancestors of modern Africans and some other archaic human species.[6] Denisovan-like mito-chondrial DNA was recently found in a 400,000 year-old 'human' bone excavated from a deep cave-shaft in northern Spain, making it look as though the ancestors of Europeans might have interbred first with Denisovans, then with Neanderthals![7] What exciting times we live in. Who knows what will turn up next?

11

Keeping it in the Family

Hybrids have always fascinated cryptozoologists and, as we shall see, they are still implicated in the creation of *les bêtes ignorées*. Hybrid appeal is nothing new. Between the thirteenth and fifteenth century fabulous creatures from the union of one or more different species inhabited all medieval bestiaries. Among the favourites were the griffon, with the body of a lion and the wings of an eagle, the leucrota, having the haunches of a stag, the breast and shins of a lion and the head of a horse, and the yale, sporting the tail of an elephant and the face of a boar.

However, contrary to what you might now think after reading the previous chapter or browsing through early manuscripts, successful hybridisation through interbreeding is actually very rare in most mammals, especially in the wild. Whereas it is, in theory anyway, comparatively uncomplicated for two closely related species to breed in captivity, the offspring are generally not as fit and healthy as their parents. In the wild, without the care and attention of the zookeepers, they would be eliminated in the face of competition from the two parent species who have had, after all, millions of years of adaptive evolution to come to terms with their environment. However in captivity, protected from this fatal competition, hybrids can thrive. Famous examples are the offspring of tigers and lions, the *liger* (lion father, tiger mother) and the *tigon* with the opposite parentage. They are healthy, indeed typically the liger is larger than either of its parents. The trouble begins when they come to breed, as the males of both hybrids have very low sperm counts, though the females are normally fertile. This follows what has become known as Haldane's rule, named after the evolutionary biologist J.B.S. Haldane who formulated it in 1922. Haldane's rule states that in a hybrid the heterogametic sex is disadvantaged by low fitness or sterility. It governs all sorts of hybrids, both plant and animal.

The mechanisms behind Haldane's rule are complex, and need not concern us here, but the consequences for ligers and tigons, not to mention theoretical hybrids between different human species, is that males (the heterogametic sex, as males have X and Y chromosomes while females have two identical X chromosomes) are usually infertile while female hybrids are not. In a *Homo neanderthalensis x sapiens* hybrid, whichever way round the parentage is arranged, the girls will have a better chance of being fertile than the boys.

Haldane's rule is not the only problem for hybrids. In all species, nuclear DNA is carried on chromosomes. While different species vary in their numbers of chromosomes, typically between ten and fifty, there is no tolerance of variation in chromosome count. One chromosome too many or one too few always leads to a serious medical condition, like Down's syndrome in humans, where sufferers carry an extra chromosome number 21. If the two parent species of a hybrid have different numbers of chromosomes, breeding is ruled out altogether as both sexes will be infertile. That female ligers and tigons can produce offspring at all is because their parents have the same numbers of chromosomes. It is not just that the lion and tiger parents are genetically fairly close, being two species in the same genus, but the equality in their chromosome count that allows the hybrids to breed. If the parental chromosome numbers of a hybrid are different, then it will be infertile. The most famous example is the mule, a hybrid between a horse and a donkey, each of which have different numbers of chromosomes. Although perfectly fit and healthy themselves, mules cannot produce viable germ cells – that is, eggs or sperm. The reason here is that the hybrid mule has an odd number of chromosomes. In this situation, any germ cells that are formed will have either one too many or one too few chromosomes. What generally happens is that the germ cells give up trying to sort this out and fail to form at all.

When it comes to humans and the *sapiens x neanderthalensis* hybrids that the DNA tells us have introduced the Neanderthal component into modern European genomes, both parents must have the same chromosome count. Although we and our Neanderthal cousins are far more closely related than tigers and lions, chromosomal compatibility does not necessarily follow from this evolutionary proximity. Our nearest primate relatives, chimpanzees, gorillas and orang-utans, have one extra pair of chromosomes compared to humans because, at some point in our evolution after we split from the great apes, two ancestral chromosomes fused to become our chromosome number 2. This chromosome imbalance is the reason why a chimp x human hybrid, the so far only theoretical *humanzee*, would certainly be infertile. Like the mule, a humanzee would not form sperm or eggs. Until very recently we did not know where on the tree of human evolution this chromosome fusion occurred. If it was during the last half million years, that is *after* humans and Neanderthal last shared a common ancestor, the two human species would have different chromosome counts and any hybrids would be infertile, like the mule. If the chromosome fusion occurred *before* the split between the two human lines, then both *H. sapiens* and *H. neanderthalensis* would have the same number of chromosomes and fertile hybrids would not be ruled out by numerical incompatibility. But what is the answer? The best way to find out is to look at the chromosomes under a microscope, but to do that requires having live cells which is, of course, impossible with Neanderthals – or at least it is until one is found alive.

Of more immediate importance to our own interest in the question of hybridisation is that the high-quality Denisovan genome sequence contained information about the chromosome number. Chromosomes are essentially very long linear strands of DNA made up of only four chemical units abbreviated A, T, C and G. DNA is a code which conveys instructions on how to build and run an organism from one generation to the next. As in any code, like a word, it is not so much the letters themselves but the order in which they occur that matters. Although the DNA alphabet has only four letters, the possible combinations are almost infinite. The sequence is all. At both ends of human and primate chromosomes there is a stretch of DNA with the sequence GGGGTT. When the ends of the two primate chromosomes fused to form human chromosome 2, these sequences of GGGGTT met head to head at the join to create the sequence GGGGTTTTGGGG. This joining segment has remained in the genome of Homo sapiens ever since. A search of the Denisovan genome found these head-to-head fragments, presumably from the fused chromosome 2, but the same search in the chimpanzee genome, where the chromosomes are still separate, found none. While perhaps not quite as conclusive as counting the chromosomes of living cells under a microscope, it is pretty good evidence that the ancestral chromosome fusion had already occurred by the time the Denisovans appeared on the scene. As that was probably half a million years before the Neanderthals, it looks as though the barrier of numerical incompatibility between the different hominids had never been erected. We were all free to breed with each other and live to see our daughters at least, remembering Haldane's rule, produce healthy grandchildren.

Even if hybrids between ourselves and our great ape cousins

would not be fertile, would they ever be conceived, let alone born live? Bernard Heuvelmans was especially fascinated by the prospect, as I discovered in his archive in Lausanne when I looked through his bulging box files of press cuttings and scientific papers on the topic. One file, coloured pink and intriguingly entitled 'Hybrides: Vrais, présumes et fabuleux' (Hybrids: True, presumed and fabulous), contained a wide range of material from, at one extreme, academic papers, such as Richard Van Gelder's essay on the classification of genera and species written for the American Museum of Natural History in New York[1] to, at the other, a tabloid French magazine's coverage of 'Queen Kong: The Liberated Lady Gorilla' illustrated by a picture of the firmly-bosomed pongid, hair swept back under a golden hairband, perched on top of a skyscraper under attack by fighter planes while clasping a hapless man in her giant hands.

In between these two extremes were cuttings of newspaper reports of actual human-ape hybrids, the most famous of which was Oliver. But before we come to him, I was astonished to come across an article in Heuvelmans' files written in 1908 by one Herman Bernolet-Moens and titled 'Experimental Researches about the descent of Man'.[2] Bernolet-Moens was a Dutch amateur scientist and scholar who, in this article, announced that he was on his way to the French Congo, which at the time covered the present-day Republic of the Congo, Gabon and the Central African Republic, in order to undertake experiments into human evolution. He had, so he said, the full support of the French government in this endeavour.

In one section of the article called 'The Artificial Fecundation of Mature Female of the Anthropoid Apes with the Sperm of Man' he made his intentions abundantly clear, stating the 'gorilla and chimpanzee will be especially fecundated with negro sperm'. In a space reserved for the anticipated results he placed a large question mark and the undertaking that 'when my work in the Congo shall be over I will substitute the result of my experiments for this point of interrogation'. This intriguing publication was a declaration of intent, a prospectus with an invitation for readers to contribute. It finished with an appeal: '. . . I trust to be favoured with the help of my readers who feel sympathy for my enterprise' adding that he had already received a donation from none other than Her Majesty the Queen of the Netherlands and from other members of the Dutch royal family. As far as I have been able to discover, Bernolet-Moens never did fill in the space reserved by the question mark. In fact, it's unclear whether he even began what to us now seems an abhorrent experiment.

Bernolet-Moens was later involved in a scandal when, in 1919, he was indicted for taking photographs of African-American schoolchildren in Washington DC, ostensibly to progress his anatomical studies. From being admired as a scholar his reputation rapidly plummeted to revulsion as a suspected paedophile. With time public perception of Bernolet-Moens swung back to respectability and he was one of the first to promote inter-racial mixing as a way to improve the prospects for our species. This view was in complete contrast to the prevailing eugenic philosophy, which we would now see as rabidly racist, which was to discourage interbreeding across ethnic boundaries and to eliminate 'inferior' types. Just this philosophy was later used to justify the atrocities of the Third Reich.

Even if Bernolet-Moens experiments in the Congo never materialised, there are a number of travellers' tales of half-ape, half-human intermediates from Africa all meticulously catalogued in Heuvelmans' files. For example, on 6 June 1926, the South African *Sunday Times* printed the following account from the prospector Paul de Chaillu who was camping near a village in the Congolese province of Katanga.

One day an extraordinary individual appeared. I say 'individual' but he looked more like a gorilla than a human. He stood about 5ft. 9in. in height judging from my own height. His legs were slightly thicker than the ordinary native's, and his arms were a good deal stronger. His body was covered in hair. But it was not as thick as that of a baboon, while his head much resembled that of the very intelligent apes one sees in zoos; yet there was a very human expression about it, too. It was a half-native type of countenance, with protruding jaw, and low receding forehead: neither man nor monkey. The nose was flat; the normal type of native nose. He had startling black eyes, brighter and more searching than those of a native. His hair was similar to a native's, except that it was longer.

When de Chaillu enquired which tribe the unusual individual belonged to, he was told that it was 'the offspring of a native woman and a gorilla father. He had come to the tribe as a boy; had just wandered in, as it were, and no one knew whence he came. He had, however, lived with the tribe since the day of his arrival, and had in time acquired all the tribal ways. I spoke through the interpreter to the chief, who informed me that he had seen similar specimens in the forest, and thought that there were many such crosses in gorilla country.'

I later came across the same story in Hedley Chilvers' book *The Seven Lost Trails of Africa*, which puts the year of de Chaillu's encounter as 1913 and adds a comment by the anthropologist Raymond Dart. Professor Dart, based in Johannesburg, who was the first to describe the species *Australopithecus afarensis*, is quoted as saying that there was nothing impossible in the abduction and impregnation of a human female by a male gorilla leading, by implication to the birth of a hybrid child.[3]

Somewhat less credible is the story carried in the *Kansas Daily Globe* of 7 July 1921 that described the elopement of a beautiful New York socialite with a strange-looking man who had come from South Africa to live in the Eastern US. He was described as being large and thickset with a muscular and powerful frame, and arms that hung down to his knees. The girl's brother, alarmed at his sister's choice of fiancé, confronted the man who broke down and explained that his mother had been abducted by a gorilla on the East Coast of Africa and that he was the result.

A brother's concern for the fate of his sister turned to revenge in another story carried by the *Kansas Daily Globe*. The heroine of the tale had been abducted by a gorilla while a tourist in Africa, carried into the forest and forced to live with the creature for a month. She somehow managed to escape and described her captor as having two toes missing from his left foot. With this key feature in mind, the girl's brother, a millionaire apparently, set out for Africa with a party of expert marksmen intent on hunting down and killing the three-toed violator of his sister's chastity. Unfortunately the *Globe* did not report what happened when, and if, the vengeful sibling confronted the malefactor.

Other pongids with amorous intent towards human females have also faced the wrath of their protectors. The Massachusetts-based *Middlesex News* of 13 February 1991 carried one such story under the headline 'Killed for a Kiss'. An orang-utan grabbed an Indonesian woman as she was undressing to bathe in a river. She screamed and fainted, and the ape ran off. Even so, it was enough for one of the villagers to track the orang-utan and kill it, despite its legally protected status. Recently I was told by a primatologist that rape by orang-utans is a known occupational hazard for female field workers in Sumatra but she did not know of any records of offspring.

In December 1980 a press report from the London *Times* correspondent in Peking (Beijing) quotes an announcement in the Shanghai newspaper *Wen Aui Bao* that Chinese authorities were considering renewing a breeding programme involving humans and chimpanzees to 'found a strain of helots for economic and technical purposes'. According to the newspaper, thirteen years previously a female chimpanzee in a medical research laboratory became pregnant after being inseminated with a man's sperm. The laboratory was later smashed up by the Red Guard and the chimp died. Mr Qi Yongxiang, the man behind the new project, was quoted as saying that the proposed humanzees would be able to drive a car, protect forests and be used for space exploration. Mr Qi's grip on the ethics of the proposal appeared a little slack. When asked whether creating such a hybrid was unethical he replied that 'semen was of no account once it had left the body and could be disposed of like manure. The hybrid would be classed as an animal, so there need be no qualms about killing it when necessary.'

An equally ambitious, and chilling, series of experiments to create ape-human hybrids was started in the early years of the Soviet Union, but only came to light after Soviet archives were opened in the 1990s.[4] The instigator was Ilya Ivanovich Ivanov, a gifted and respected zoologist and an early pioneer of artificial insemination as an aid to improved efficiency in the breeding of horses. He announced his intentions to create human–chimpanzee hybrids at the 1910 World Congress of Zoology in Graz, Austria but only after the 1917 revolution was he able to attract official backing. His plans received enthusiastic backing from Bolshevik intellectuals who saw the opportunity for anti-religious propaganda in the project. There was also support from Americans eager for a practical demonstration of Darwin's evolutionary theory, now widely accepted, that humans were descended from apes. Ivanov was given a large grant to set up a research facility at Kindia in French Guinea in West Africa. There he set about capturing chimpanzees and to arrange for the insemination of females by human sperm and, for the reverse cross, inseminate local women with ape sperm in exchange for payment. He did succeed in capturing a number of chimpanzees but found no volunteers among the local population for insemination by ape sperm. He explained this reluctance down to their fear of being ostracised by the community, as was the fate of women that had been raped by chimpanzees in the wild. He did not give up and planned to inseminate women without their consent during faked medical examinations for other purposes but this was stopped by the governor. With no pregnancies among the captive female chimps, Ivanov moved back to Russia to continue his experiments at a specially constructed research facility at Sukhumi in Abkhazia. He took twenty chimpanzees with him, but only four survived the trip.

He did find five female volunteers willing to be inseminated

but was rapidly running out of male apes to provide the necessary sperm and no pregnancies resulted. Ivanov persisted until
1930 but the mood of his political supporters shifted and he
was arrested and exiled to Kazakhstan where he died two years
later.

I did not find any material specifically relating to the Ivanov
experiments among Heuvelmans' papers, but in a book written
with Boris Porchnev he tells the story of a Russian doctor who
claimed he knew of a Soviet concentration camp where a new
race of man was being created by hybridisation between humans
and apes.[5]

Let me now return to the hybrid story which occupied most
space in Heuvelmans' pink folder, that of Oliver. He was born
in the Congo in about 1968 but spent most of his life in America
until he died, aged around thirty, at a retirement home for circus
and medical research chimps near San Antonio, Texas. Unlike
most chimpanzees who habitually walk on both arms and legs,
Oliver was consistently bipedal. He was intended to be part of
a well-known touring animal act with three other chimps brought
over from Africa at the same time. But the other chimps wanted
nothing to do with Oliver, so he was left out of the shows.
According to his owners, Frank and Janet Burger, Oliver was
much brighter than their other apes and was soon helping around
the house. 'You could send him on chores. He would take the
wheelbarrow and empty the hay and straw from the stalls. And
when it was time to feed the dogs, he would get the pans and
mix the dog food for me. I'd get it ready and he'd mix it,' Janet
said.

As Oliver grew older he adopted some more human habits.
'This guy Oliver, he enjoyed sitting down at night and having
a drink and watching television. He'd mix his own. He'd put a
shot of whiskey and put some Seven-Up in there, stir it and
drink it.' Oliver looked different too, with a smaller head, shorter
arms and less hair than a regular chimp. He had a squarer
face, and a more human-like expression than other chimps.
The Burgers were forced to sell when Oliver was approaching

sexual maturity. As Mrs Burger explained to reporters: 'He had sex on his mind. The old hormones flared up. He didn't care about the female chimps we had, he started trying to have sex with me and any other woman.' Oliver's next owner was a New York City lawyer who saw his new charge's earning potential. He took Oliver on tour all over the world and indeed, during his period of ownership, tried to pass him off as a Bigfoot.

What was Oliver? Was he just an unusual chimpanzee, or was he really a human–chimp hybrid from the Congo? Recalling the difference in the chromosome number between humans and chimps, it should have been an easy thing to establish. All that was needed was to grow some of Oliver's cells in a culture dish and look at them under a microscope when they were about to divide. At this stage, chromosomes are at their most condensed and can be individually recognised and counted. Normal chimpanzees have forty-eight chromosomes and humans only forty-six because of the ancestral fusion of two chromosomes to form human chromosome 2, as we saw earlier in this chapter. According to reports in *The Sunday Times* of 5 Sept 1976, the testing was carried out at the University of Chicago and counted forty-seven chromosomes in Oliver's cells, the number expected in a chimp–human hybrid. If these reports were accurate, Oliver may well have been a genuine humanzee.

The same *Sunday Times* report revealed a plan to breed from Oliver, carried under the headline 'Girl plans to mate with Ape'. The proposed encounter was to take place in a Tokyo hotel after Oliver and Miss Hiroko Tagawa, a nineteen-year-old actress temporarily working in a sushi restaurant, had been properly introduced.

Though I have been unable to find confirmation that their anticipated conjugation ever took place, I have been able to find out more about the background to Oliver's chromosome count. According to other press reports in the Heuvelmans archive, Oliver's cells were to be sent to Dr David H. Ledbetter, a geneticist at the University of Chicago. During my time in medical genetics,

I had come across a Dr Ledbetter and knew his work on a rare genetic disorder called Fragile X syndrome, whose unusual causation became a key part of my genetics lectures to the Oxford medical students.

I tracked Dr Ledbetter down through a former colleague in Chicago and got a swift reply that he was indeed the man who had examined Oliver's cells. He had grown them on a cell-culture slide and counted the chromosomes as they condensed during cell division. Oliver did not have forty-seven chromosomes, as widely reported, but the regular forty-eight. In Dr Ledbetter's words, he was 'just a chimp that liked to stand up'. He also told me that he had intended to write this investigation up as a paper, but never got round to it. He felt, as I would have done in the circumstances, that it was a negative result of no importance that was unlikely ever to be published in a serious journal. The press meanwhile completely lost interest in Oliver on hearing that he was a regular chimp, but the myth that his hybrid status had been confirmed by chromosome analysis has continued.

Heuvelmans' bulging pink file held many more tales of hybrids between animals of all descriptions. They must have held a particular fascination for the father of cryptozoology, a fascination that for many of his followers continues to this day, as we shall see. But let us close the file at this point and return to our own species.

One point that has so far been overlooked in the excitement of finding significant chunks of Neanderthal and Denisovan DNA in our genomes, with the promise of more to come is this. Where does it leave us, *Homo sapiens*, as a species? The biological definition insists that a species can only breed successfully in the wild with itself. It is an important evolutionary principle for the Darwinian concept of the origin of species, which requires that at some point there must be some sort of barrier to interbreeding. Often this barrier is geographical. There are many examples, including Darwin's original work on the finches living on the scattered islands of the Galapagos archipelago. It was through his observations of these birds that Darwin consolidated and developed his ideas on the origin of species through natural selection. In the Galapagos Darwin noticed that the finches looked quite different on each island. In fact he thought they were completely different birds altogether until the ornithologist John Gould examined the specimens Darwin had brought back with him and pointed out that they were all closely related. Darwin eventually realised that these distinguishing features may have evolved through isolation, something that would not have happened had the finches continually interbred with birds from other islands.

Because of Darwin's somewhat delayed insight into the Galapagos finches, and hundreds of examples since that time, species definition has become very strict. Speciation, the process of creating new species, in populations of wild animals requires that something happens to make interbreeding between them impossible, a good example being the chromosome fusion in the human ancestral line that prevents interbreeding in the wild between humans and chimps. It follows that if what are thought of as two different species of animals do successfully interbreed in the wild and produce healthy fertile offspring that are able to survive, then they are not strictly-speaking two species but one.

Where does that leave us? Collateral human species, like the Neanderthals, have traditionally been defined by their appearance deduced from fossils. However, it is quite impossible to tell from the appearance of the skeleton whether Neanderthals were a different species from *Homo sapiens* by the strict biological definition of genetic incompatibility. But now all that has changed. We now believe that Neanderthals, Denisovans and our own *Homo sapiens* did interbreed both successfully and in the wild. If the offspring of the occasional inter-species human x chimp liaisons, possibly like the unusual individual encountered by Paul de Chaillu in the Congo, chromosome incompatibility would have prevented the parental DNA being passed on to the next generation and we would not have been able to detect it in modern humans. But with the Neanderthals and Denisovans, we can. So, to my mind, this means that by the strictly applied biological criteria insisted on by modern taxonomy, we are all, Neanderthals, Denisovans and the rest that have left a trace of their DNA in our genomes, members of the same species. While this will appear to many to be a semantic distinction, there is one very practical implication. While it is not illegal to kill an unknown species or indeed a hybrid, if we are all in the same species then there can be no hybrids. By this reasoning, to kill a Bigfoot, a yeti, a Neanderthal or a Denisovan, as well as being a travesty, might also be classed as homicide.

12

The Crimson Casket

There were three of us at the table. The waiter recommended the Chablis, fetched a bottle and poured out two glasses. Lady Antonia and I savoured the bouquet, took an approving sip and raised a glass to my other guest, the small crimson leather casket lying in the middle of the table. Harold had always disapproved of this place. He had always hated ties and only tolerated wearing them at funerals (black) and for watching cricket at Lord's (orange and yellow, the colours of the MCC). That gentlemen must wear ties was the club's unwavering rule. So, despite numerous invitations, Harold rarely came to the Athenaeum during his lifetime. Today he had no choice. You see, Harold had died two years before and the crimson casket in the middle of the table contained a lock of his hair, within which were hidden a few molecules of his biological essence, his DNA. It might have been a scene from one of his own plays.

After our lunch, Lady Antonia opened the casket and took out the small plastic wallet within which were curled twenty or so dark brown hairs. I took the packet and examined it. Inside was a handwritten note 'HP 10 Jan 2002 – cut by Delilah. AF'. Lady Antonia explained that she had cut the lock from her late husband's head just before he went into hospital for his first course of chemotherapy to treat a recently diagnosed cancer. Harold Pinter died six years later in 2008.

A few weeks before our meeting, Lady Antonia Fraser, Pinter's widow, had asked me during a chance meeting whether, even after his death, I could link Harold to one of the seven maternal European clan mothers that I had identified and who became the heroines of my first book, *The Seven Daughters of Eve*. That would have been an easy task had Pinter been alive. A simple mitochondrial DNA analysis from a cheek swab would have done it. I was curious to see if I could still fulfil her request, knowing that all I had to work with was a few hair shafts that were last truly alive ten years before. Until this time I had only ever tried to retrieve DNA from hair roots, the follicles containing thousands of cells. Hair shafts, extruded by the follicles, are strictly speaking dead, but I was aware that new forensic techniques had often managed to find traces of DNA even within these lifeless strands.

Rather than attempt to apply these demanding new techniques in my own laboratory, I called one of my former colleagues, Dr Terry Melton. From her time in my laboratory, where she was working on genetic links between Polynesians and the indigenous people of Taiwan, I knew Terry to be an extremely careful and professional scientist. I also knew that after she left my lab, Terry had set up a company specialising in the forensic genetic analysis of hair samples. By the time I spoke to her, she had built up a good reputation among the forensic community with all the necessary accreditations and a string of clients from law enforcement agencies around the world. I asked Terry whether she would take on the task of analysing Harold Pinter's hair and she agreed.

Though I had twenty strands of Pinter's hair, Terry only needed two, and even then the second hair was only a backup in case the first analysis failed. The lab report gave me a DNA sequence that identified Pinter as a member of the clan of Katrine. The fine details of the sequence were enough to place his ancestry somewhere in the Ukraine and to show that his genetic finger-print was of a type commonly found among Ashkenazi Jews. When I relayed the results to Lady Antonia, she told me that on his mother's side Harold had indeed come from a Jewish family from Kiev.

That was a fantastic outcome in itself. Lady Antonia was delighted to have proof of her late husband's connection to his Eastern European roots. It also meant that the technology was now advanced to a point where I could expect to recover DNA from hair shafts alone, with no need for roots, as well as in samples that were not fresh but several years old. Just as impor-tantly, the experiment with Harold's hair had shown that it was possible using this protocol completely to remove surface contamination. At our meeting, I had taken a DNA sample from Lady Antonia Fraser and found she was a descendant of Jasmine, one of the other European clan mothers. Though she had cut and handled Harold's hair herself, there was no trace of Taran DNA in her husband's sample after the forensic clean-up.

My analysis of Harold Pinter's hair had been carried out with no thoughts of yetis. When I did begin to think seriously about the current project I was immediately aware of the possibilities the Pinter result opened up for identifying hairs that had been attributed to anomalous primates. While I may have been lucky enough to obtain a fresh specimen from a current expedition, I reckoned that most yeti and Bigfoot hairs were already in museums or private collections. They had probably been handled, so were almost certain to have human DNA contamination on their surface. That was only to be expected, even with fresh samples. The results from Pinter's hair, which was a decade old when I tested it in 2012, promised that there was a good chance of success with much older material and, crucially, that the curse of contamination could be overcome.

Even so, I wanted to be quite sure that the protocol that Terry had used on Pinter's lock also worked on animal hair. To prove that, I began to look out for suitable material to test. I soon found what I wanted in the University Museum of Natural History in Oxford. Just inside the door stands a stuffed Shetland pony. Its label told me that 'Mandy' had died in a Yorkshire zoo in 1980, over thirty years before. More significantly, the same label also invited visitors to touch and stroke the pony, and even as I looked on a group of schoolchildren were doing just that. Mandy was the perfect specimen for my purpose, decades old and coated in human DNA from hundreds, if not thousands, of schoolchildren. If I could recover DNA from this sample and, despite the human surface contamination, correctly identify it as coming from a pony, then I would feel confident enough to launch an appeal to museums and cryptozoologists worldwide to send me a little of what they had.

It proved to be even easier than I had thought. From just a two-centimetre length of a single hair shaft, Terry managed to extract enough DNA to generate a fingerprint which, when compared to the thousands now available on published databases, came back with just one unequivocal match: *Equus caballus*, the domesticated horse. There was, as I had hoped, absolutely no sign of any DNA from the generations of children who had stroked Mandy's flowing mane.

PART II

13

The DNA Toolbox

I had identified Harold Pinter's maternal origins from the fragments of a special sort of DNA with which I was very familiar. A great deal of my research on human origins over the last twenty-five years has been based on interpreting the echoes of our past that still resonate in mitochondrial DNA. As the name suggests, this DNA is found inside the mitochondria, which are tiny structures lying between the nucleus and the membrane of every cell. Mitochondria play a crucial part in aerobic metabolism by releasing energy to the cell. This is no minor cameo role in the theatre of life. A few milligrams of cyanide completely shuts down the mitochondria and death follows in minutes.

Mitochondrial DNA is far and away my favourite piece of DNA. Over the last twenty years it has helped me to discover the ancestral origins of, among others, Polynesians, ancient Europeans and the British. It has always been my first choice when attempting to recover DNA from human fossils or, as I anticipated would be the case for old yeti hairs, any sample where there was likely to be very little DNA remaining.

Mitochondrial DNA is always inherited down the maternal line. We get it from our mothers, who got it from theirs, who got it from theirs and so on back into deep time. It is also the only DNA to survive in sufficient quantity for routine analysis of a single hair shaft using current technology. The other kind of DNA in our cells is confined to the long chromosomes found in the nucleus. Although this nuclear DNA contains almost all the genes for the proper functioning of our bodies, we inherit it from each of our parents, not just our mothers as we do mitochondrial DNA. This complicates the interpretation, especially when you consider that our parents inherited their nuclear DNA from all four of our grandparents. To complicate matters even more, the number of ancestors contributing to our own nuclear DNA doubles for each generation as we go back further into the past.

With mitochondrial DNA things are much simpler. However far back we choose to go, hundreds, thousands, even hundreds of thousands of years, there was only ever one person, one woman, in that generation who was our direct maternal ancestor. All our mitochondrial DNA came from her and I like to think she is with us still with every breath we take. The pattern of inheritance certainly keeps thing simple, even though it means that mitochondrial DNA will only tell us about this one matrilineal genealogy. I have written extensively about this in earlier books, especially *The Seven Daughters of Eve*, so I won't repeat myself here, except to say that the simplicity of mitochondrial inheritance has also been invaluable to my research for *The Nature of the Beast*.

The feature of mitochondrial DNA that makes it the first choice for extractions from human fossils is its relative abundance. There are a lot more copies of mitochondrial DNA than nuclear DNA in each cell and as the main function of mitochondria is providing energy, active cells like muscle and nerve contain the most mitochondria. But even relatively docile cells have far more copies of mitochondrial than nuclear genes. This abundance of mitochondrial DNA makes it the preferred target in situations, like fossils, where there isn't much DNA to work on. The same is true of hair shafts, which contain only a small amount of DNA left over from the active follicles from which they grow. Though there were times during this project, as you will see, when I did need to study nuclear DNA, the work on species identification from hair shafts relied on mitochondrial DNA, my old favourite, throughout.

As I have explained, most of my research over the years has been on human DNA and what it can tell us about our past. Mitochondrial DNA has been very useful partly because of its uncomplicated matrilineal inheritance pattern, partly because there is a lot of it in a cell but also because it is carried from generation to generation virtually unchanged. Virtually, but not completely. The random process of mutation introduces tiny changes into the mitochondrial DNA sequence, but at an extremely low rate.

The segment of mitochondrial DNA that I have used for most of my population origin work is known as the 'control region' because it is in overall charge of making sure the mitochondrial DNA functions correctly. Within the control region, which is about one thousand DNA bases long, is a four hundred base segment given the name Hyper-Variable Region 1 – or HVS1 for short. With HVS1 mutations happening on average only once in every twenty thousand years the name flatters its mutability, but it is the case that HVS1 changes much faster than the rest of the mitochondrial DNA, for reasons we will come to shortly.

This admittedly languorous mutation rate is in fact ideal for exploring the main events in human evolution since we started out as a separate species, about two hundred thousand years ago. One woman alive at the time was the mitochondrial ancestor of everyone now living on the planet. Unsurprisingly she became known as 'Mitochondrial Eve' and as far as we are able to tell she lived on the plains of East Africa. Her body has never been found, and she was unremarkable. Like the other women, and men, who lived at the same time, and there are estimated to have been about ten thousand of them, Eve and her contemporaries were hunter-gatherers. They operated as a group of around twenty and worked together to ambush the same prey animals that still live in the savannah. Of all peoples alive today, Eve most closely resembled the fast disappearing hunter-gatherers of the Kalahari. The only thing that distinguishes Eve from the other women around her was that she was the only one to have matrilineal descendants who are alive today. All of us are Eve's descendants. The other women around, or their matrilineal descendants, either didn't have any children, or had only sons. Either way, their mitochondrial DNA disappeared at some point during the intervening millennia between then and now.

Comparing the mutations in people from different parts of the world, scientists like me have been able to link individuals and populations together to discover how they are related through their shared maternal ancestry and from that get an idea of how our ancestors spread out from Africa to colonise the world.

That we are all descended through the female line from just one woman, 'Mitochondrial Eve', is astonishing enough, but it immediately raises the question of Eve's own ancestors. Who did she get her mitochondrial DNA from? From her own mother, of course, and her mother before that. But the maternal line stretches further and further back in time until Eve's matrilineal ancestor is not even human, well not *Homo sapiens* anyway. Eve's distant ancestors will have had genetic relatives with other types of human, including Neanderthals perhaps. Since we now know the DNA sequence of Neanderthal mitochondria from fossil remains, the link between Eve and these other types of human can be compared.

Eventually, by tracking further and further back, we reach an ancestor who is shared by both *Homo sapiens* and Neanderthals. And then, further back still, to an ancestor shared by all humans and our cousins, the great apes – chimpanzee, gorilla and orang-utan. These relations can also be explored using mitochondrial DNA and show that, among the great apes, we are most closely related to chimpanzees, then to gorillas and, most distantly, to orang-utans. Further back still we can trace our common ancestry to other primates like gibbons, lemurs, marmosets and baboons and, beyond that, to all other mammals.

Even though the mutation rate of the mitochondrial control region (by which I mean the HVS1 segment) is very slow, changing roughly only once every twenty thousand years, this is still too rapid for investigating deeper links of the order of millions of years. There are just too many differences to deal with. While this is fine for working on modern humans, and for our other human cousins like Neanderthals, it starts to get more complicated when the links we want to investigate are between ourselves and the great apes, and becomes virtually impossible when we go further back to connect to other primates and the rest of the animal kingdom.

Fortunately there is another stretch of mitochondrial DNA with a much slower mutation rate that is better suited for these situations. Let me explain. The reason that the mutation rate of the control region is relatively fast compared to the rest of mitochondrial DNA is that the exact DNA sequence doesn't appear to matter. Mutations happen at the same rate all round mitochondrial DNA, but usually disrupt the function of the genes in which they occur. There is no room for slack in DNA with such a vital function as aerobic metabolism, so any individual whose mitochondrial DNA gets hit by a random mutation in an active gene will die very fast, or more likely will never be born at all. The relentless process of natural selection soon weeds out individuals with malfunctioning mitochondria. The majority of mutations that occur outside the control region simply do not get passed on. However, within the control region mutations are far less harmful to mitochondrial function. It seems that, while the mitochondria need a segment of DNA to occupy the control region, the precise sequence of this segment doesn't seem to matter. The mutations are not eliminated and do get passed on from one generation to the next. Control region mutations have accumulated in countless matrilineal genealogies over the last hundred thousand years since Mitochondrial Eve. As a result there are tens of thousands of different mitochondrial DNAs throughout the world that can be used to explore the links between modern humans and to discover where different people came from.

Outside the control region, it is a very different story. Thanks to the weeding out process of natural selection, hardly any mutations survive to be passed on. In the most stable of these DNA segments all living humans have the same sequence. Indeed even our cousins, the Neanderthals, have the same sequence as ourselves. This segment is part of a gene called 12S RNA; it helps mitochondria assemble the enzymes needed for aerobic metabolism. RNA is the acronym for ribonucleic acid and 12S denotes its size, though this is not relevant to us. The reason I chose mitochondrial 12S RNA for the species identification of hair samples attributed to the yeti and Bigfoot is because its sequence is known for all known mammals. Importantly for us, these sequences have been deposited in the public access database GenBank and are available for comparison. From my point of view, this was ideal. Terry Melton and I had managed to recover mitochondrial control region sequences from Harold Pinter's hair shafts. If we could do the same with mitochondrial 12S RNA sequences from the hair samples from anomalous primates that I hoped to collect, and compare them with what was in GenBank, then that should either identify the species from which the sample derived with absolute precision or tell us that it was from a hitherto unknown species.

In past species identification experiments with difficult samples it was necessary to design the analysis around the likeliest suspects. For example, for the *migoi* hairs from Bhutan that my student Helen Chandler worked on in 2000, we tailored the experiments to detect only bear DNA. This was in order to avoid picking up the contamination with human DNA that plagues the analysis of every sample. It worked, at least in two out of the three samples, because these samples were indeed two different species of bear, as we had suspected. Now, using Terry Melton's protocol, we had a way of getting rid of all human contamination from the hair samples. We would no longer have to guess the likely species origin of a sample before we started to analyse it in order to avoid detecting instead the contaminating human DNA on the surface. Now we could design the analysis to detect and sequence mitochondrial 12S RNA from any mammal without the need for prior knowledge of its identity. This was clearly important for all our samples, whether from hairs belonging to known animals that had been misattributed to yetis or Bigfoot or from any unknown species that we might encounter.

For any sequence that we recovered from a hair sample, if there were an exact, or very close, mitochondrial 12S RNA match in GenBank, the source species for the specimen will have been unequivocally identified. Alternatively, if the specimen were from a new species, which by definition means that it will not be in the database, then its position on the evolutionary tree relative to known species could be deduced. For example, if the specimen were from an anomalous primate that was neither a human nor one of the great apes, either chimpanzee, gorilla or orang-utan, its relative genetic similarity to each of them could be estimated from the 12S RNA sequence, even though there would be no exact matches. Equally, a new species of ape or monkey or bear could be identified by similar means.

For cases where the mitochondrial 12S RNA sequence from a hair sample identified it as human, I planned to use the much more variable control region sequence to distinguish *Homo sapiens* from *Homo neanderthalensis*. The 12S RNA sequences for both these species are the same but, as we have seen, they are very different in the much more variable control region. As well as distinguishing *Homo sapiens* from other hominids, the control region sequence would also help identify the approximate geographical origin of any modern human hairs that I was given to test.

Anticipating that many of the samples would be in bad condition, either through age or for some other reason, I decided to use an experimental protocol that treated all samples as if they were ancient DNA, the sort I had been used to getting from fossils. Ancient DNA is broken down into short segments over time and the sequence of even a hundred bases has to be built up from smaller pieces. This puts a practical limit on the length of the 12S RNA segment that could be routinely sequenced from an old hair sample. I also decided to concentrate on a segment of 104 bases in the mitochondrial 12S RNA gene. This was not so long that it would be hard to analyse in a degraded sample, and not too short, I hoped, to miss mutations that would identify the species.

To cut a long story short, I asked Terry if she would carry out the hair shaft analysis on my anomalous primate samples and she agreed. For me, there was another advantage to the arrangement beyond the knowledge that I could absolutely rely on the results from Terry's lab, in that she was accustomed to working with mitochondrial 12S RNA. In the very controversial field of yeti and Bigfoot research, I anticipated that my results might be challenged at some stage, more so than in my regular work. To have an independent lab carry out the analysis and to receive a report of a standard that could stand up in court was a great advantage.

But before we start in earnest on how I applied these new techniques to the cryptozoological material, let us have a look at the few relevant genetic studies that have already been undertaken.

14

Good Science, Bad Science

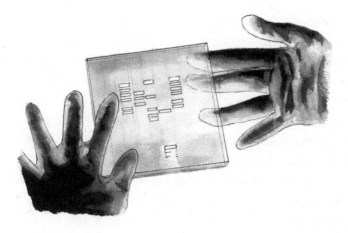

One of the surprises that greeted me when I first plunged into the murky waters of cryptozoology was its relationship with mainstream science. In past decades people like Bernard Heuvelmans, John Napier of the Smithsonian Institution, Grover L. Krantz of Washington State University and others had spent a great deal of time trying to identify the creatures they were dealing with. These men were trained as scientists and by and large kept to the principles of their calling. They may not have succeeded in providing the key proofs that would satisfy a critical audience, but at least they were trying their best to do so. That much is clear, particularly in the case of Bernard Heuvelmans, as was very evident from his correspondence in the Lausanne archive. He may have enjoyed the celebrity that surrounded him, but in the quiet of his own study where he catalogued every press report and every letter on his subject, systematically marking each one with his red felt-tip pen, he was clearly a serious scholar. I saw no hint in his archive that he was anything less than honest.

He may, on occasions, have been less critical than he might have been, but there was no sign of deception or fabrication. He worked only slightly beyond the orbits of full-time professional scientists, like Napier, but he was on good terms with them and had their respect. At least he did until he was undone by the Minnesota Iceman.

Heuvelmans did not do any laboratory research himself, but was in touch with scientists who did have the facilities to work on material brought back from the wild. Unlike Heuvelmans, who devoted most of his life to cryptozoology, these scientists had plenty of other things to occupy them. Without proper financial support they were only willing to give a portion of their time to these fringe cryptozoology projects, and rightly so, as the income from these projects never covered their expenses. Aside from the occasional sponsorship of wealthy men like Tom Slick and his associate Kirk Johnson Sr, the scientists who worked cryptid material did so with no financial support. The only other way of getting the necessary funds to pay for the research was through writing books and articles, and appearing on the radio or TV.

Of course, there is a price to pay to please the *avidum genus auricularum*, the enduring mass of David Hume's gazing populace, which any media outlet must amuse and entertain and, if necessary, inform. While the 'inform' aspect of reportage might take priority in the coverage of major political or financial events, it is the 'amuse and entertain' potential of cryptid stories that appeals – not necessarily the ideal foundation for proper scientific enquiry. That is not to say that the 'amuse and entertain' potential of scientific research should be overlooked. I have thought for a long time that scientists do not pay enough attention to the appeal of their research to the general public who, after all, are very often their paymasters and by whom they are entrusted to work on their behalf. But support is quickly withdrawn from any scientist who deliberately invents his or her results. For a professional scientist to be found guilty of fabrication is the end of the line. They face dismissal, and sometime prosecution. The integrity of scientists is absolutely essential for their world to survive and prosper. All professional scientists understand this and ruthlessly pursue and eliminate transgressors.

The way professional standards are enforced, or at least encouraged, is by publication of results in scientific journals after close scrutiny by other scientists in a process known as peer review. In other walks of life that might seem an unsatisfactory way to ensure honesty, and perhaps it is. The peer-review system is not primarily there to root out corrupt and dishonest practices. It is there to ensure that the experiments have been properly carried out, that there is enough detail for them to be repeated and, most importantly, that the conclusions are supported by the evidence. Without peer-reviewed publication it could be argued that a piece of research is not really science at all.

That can be a harsh conclusion. Some of my own work has not been through this process yet I still consider it valid science. I have done plenty of experiments that produced negative results, and not bothered to write them up for publication. This is the fate shared by much of the cryptid work done by professional scientists. Hardly any of it has been through the peer-review system and often for very good reasons. Like the press, scientific journals prefer novelty in what they publish. The key difference between the two is that, while the media are able to, and often do, report a newsworthy project when it is in the planning stage, a scientific journal needs evidence, which can only come after the research is complete. Take, for example, the case of Oliver the chimpanzee's crucial chromosome count. The work had been done but the results were not written up for a scientific journal. The popular press, while keen to write about the proposed testing in advance, lost interest in reporting the result when it showed that Oliver was just an ordinary chimp.

I can give you an example from my own laboratory. As I have already mentioned, in 2000 I received three samples from the Himalayan Kingdom of Bhutan that were attributed to the *migoi*, the Bhutanese equivalent of the yeti. The genetic investigation was to be part of a documentary film about the *migoi* being made by Harry Marshall of Icon Films. (It was also Harry's company that made 'The Bigfoot Files' which followed the 'Yeti Enigma' project.) All three of the *migoi* samples that reached the lab had compelling provenances. The first was from a skin kept in a monastery that was used to treat sick monks. A night spent sleeping on the skin of the *migoi* was enough to dispel almost any malady, apparently. The second sample came from a tree where 'the *migoi* scratches its skin' while the third was found in a *migoi* nest inside a hollow tree to which the film crew was directed by the King of Bhutan's *migoi* steward. Like the swans of England, all Bhutanese *migois* are under royal protection.

After a frustrating few months of trial and error, my research student Helen Chandler managed to get some DNA from two of the three samples and identified the species through their sequences. The skin with remarkable healing properties came from the brown bear *Ursus arctos*, while the hair left in the traditional scratching tree was from an Asiatic black bear *Ursus thibetanus*. The third sample, a single hair from the hollow tree, we could not identify. When the film was broadcast as 'Wildman: Hunt for the Yeti' there was only a brief mention of the bear identifications, but a good deal about the unidentified third hair. Indeed in the closing frames of the film I am heard to say words to the effect that: 'Normally we would have no trouble identifying a good DNA sample. But then we have not dealt with the yeti before.' A very hammy performance I thought, when I saw it again recently.

Of course, it is perfectly understandable that, as a film-maker, Harry would want to end the film about yetis on a note of mystery, but the point is that while we could have written up the bear results, no journal would even consider publishing what we found out about the mystery hair, which was absolutely nothing. Just like Dr Ledbetter, we 'never got around' to sending the bear results to a journal, though it did produce a chapter in Helen's thesis – which is at least in the public domain in the sense that there is a copy lodged somewhere in the vaults of the Bodleian Library in Oxford. The hair that we were unable to identify soon transmuted into the hair of 'an unidentified species' and became something of a celebrity in the shadow world of cryptozoology. I don't think I have ever had as many enquiries about any of my work as about that third hair. But as you will read later, I recently tested what remained of the mystery hair and made a positive, and very surprising, identification.

When I set out on this current project, I was frustrated that so many press reports of yeti and Bigfoot samples that were 'on their way to be DNA tested' were never followed up. The results of these tests were seldom, if ever, reported. More cases of 'I didn't get around to it', I suspected, so I made it my business to ask some of these scientists if they could tell me what they had found and, if possible, let me have a look at the data.

What I found was a rather sorry state of confusion. Samples that I knew had been sent to good labs by enthusiasts were often lost or the results not relayed back to the donors. Material that had been carefully collected and submitted with high expecta-tions was often treated with casual disdain. Being on the receiving end of unsolicited sample donations, I am sometimes guilty of not acknowledging receipt right away or not being quick to return results. Cryptozoologists have sent samples for various forms of DNA testing to my lab and others, but rarely consider who is going to pay for it. Unsurprisingly, a sample that arrives uninvited and unaccompanied by the means to pay for the analysis is not always given the priority the donor considers it deserves.

You may now understand why lab results, properly written up, are such a rarity in the cryptid world. That doesn't stop references to genetic testing in the many books written around the subject. For example, in 2000 hair was recovered from the site in southern Washington where a Bigfoot was alleged to have rested on soft ground, leaving a much-studied impression: the famous Skookum cast. This hair was sent for analysis to Dr Craig Newton at the University of British Columbia. This is referred to in Dr Jeff Meldrum's book *Sasquatch: Legend Meets Science* and tells us that Dr Newton chose to analyse nuclear rather than mitochondrial DNA.[1] His conclusions, that he was unable to rule out human contamination, are quoted by Dr Meldrum but there is no reference to the data, which I presumed Dr Newton had 'never got around' to publishing. So I tracked him down through colleagues at the University of British Columbia and this is what he told me:

'At the time I was working on Kermode bear genetics (*Kermode bears live on the coasts of Alaska and British Columbia*). I tried my universal melanocortin 1 receptor primers (*like the mitochondrial 12S RNA gene that I was using, able to identify species origin*) on one or two of the Skookum cast hairs but only really managed to amplify myself. The amplifications (*the process of increasing the amount of DNA in the test-tube until there is enough to sequence*) were pushed to the limit and the DNA sequence was bad, so I didn't really believe anything, hence no publications or anything like that. The problem was old degraded hair samples to start with and targeting a nuclear gene with low copy number per sample. The only sequence I could ever read well was my own. The idea at the time was to be able to rule out false sightings – bear, moose, hairy humans etc. – assuming we had reasonably good samples to start with. But I've seen all sorts of things: apple core chewings, peanut husks, unusual turds, saliva smears off windows, bloody hair masses – but to this point nothing I've been able to get good data from. I am a bit ashamed, but then nobody was writing cheques and there is only so much effort that can be spent on such whimsies.'

I have found only two examples where the results of DNA analysis have been published in the scientific press. In 2004 Michel Milinkovitch[2] and his colleagues claimed to have tested a clump of hair brought back from Nepal by Peter Matthiessen, the explorer and author of *The Snow Leopard*.[3] The DNA results showed it to be most likely from an ungulate (a hoofed animal) of some sort, though the species was not identified. My suspicions about this paper were raised when I could find no trace of the sequence in the database into which, according to the paper, it had been accessioned. Suspicion mounted when I read that their findings confirmed Captain Haddock's hypothesis, elucidated in Hergé's

Tintin in Tibet, when Haddock addressed the yeti as 'You odd-toed ungu-late.' Only then did I notice the footnote that '. . . evolutionary biologists need to retain a sense of humour' and the publication date: 1 April. Fortunately, as it turned out, I missed the footnote on first reading and had already contacted Prof. Milinkovitch at the University of Geneva, where he runs the Laboratory of Artificial and Natural Evolution, to enquire about this research. After introducing myself and my current project I asked about the Matthiessen sample:

> I recently came across a reference to your 2004 paper in Mol. Phylo.Evol in which you sequenced the mitochondrial 12S RNA gene from a sample brought back from Nepal by Peter Matthiessen and attributed to a *mehti (a local name for the yeti).* I assumed this paper to be a good joke, as I could not find any yeti sequences in the EMBL database and you did not give accession numbers.
>
> Of course it is reassuring to have Captain Haddock's preliminary results on the identity of the creature confirmed by DNA sequencing, and (had I located the sequence) I would have seen whether EMBL or GenBank could provide even closer identification. As your paper has been referenced in all seriousness by other workers, I just thought I should double-check that it was genuine.

When I read the footnote in the journal, I immediately emailed Prof. Milinkovitch to say that now I knew it was a practical joke, there was no need for him to reply. But reply he did, which is when I got the whole story. Prof. Milinkovitch really had tested the *mehti* hair brought back by Matthiessen. It had arrived in the lab because Matthiessen had known the distinguished evolutionary geneticist Dr George Amato, now the Director of the Sackler Institute for Comparative Genomics at the American Museum of Natural History in New York, and had asked for his help in doing a DNA analysis on the *mehti* hair sample.

As Matthiessen describes in *The Snow Leopard*, he was on an expedition to the Himalayas in 1992 when he and his companions came upon some unusual footprints in the snow. These were immediately identified by his guides as the prints of a *mehti*. A skein of twisted hair was found beside a river at the bottom of a nearby gorge and this was also identified as belonging to a *mehti* by their Nepalese guides.

At first Prof. Milinkovitch thought that Matthiessen was pulling their legs. But he was deadly serious. He was convinced he had a genuine *mehti* specimen. Correctly, Matthiessen insisted that the identification of any new species would have to be reported to the Nepalese government before publication, a request with which Milinkovitch and Amato were happy to comply. In the event, and to Matthiessen's bitter disappointment, the DNA analysis clearly showed that the *mehti* hair sample was actually from a horse, *Equus caballus*. The published paper did not reveal the species identity, only that it was an ungulate and that this confirmed Captain Haddock's earlier conclusion on the identity of the yeti.

As Prof. Milinkovitch added to his reply: 'I am Belgian, and as any real Belgian guy, I know *Tintin au Tibet* by heart.'

The *mehti* sequence, though declined by GenBank, is publicly accessible through the professor's laboratory website.[4] I ran a 150 base segment through a GenBank search to see if there were other species identifications. Ninety-eight of the first hundred sequence matches came back as *Equus caballus*, the horse. However, the last two were matches with the kiang and the onager respectively. Both are wild asses. While the onager is widespread in India, Pakistan and the Middle East, the kiang is found in Tibet and occasionally over the border in Ladakh, India and northern Nepal. Although Matthiessen's *mehti* was probably just an ordinary horse, it might just have been a kiang. It most certainly was not a primate.

The second example, published in the respected *Trends in Ecology and Evolution*, reported DNA results from a sasquatch sighting in the town of Teslin, Yukon.[5] Nine residents witnessed a large biped moving through the brush outside the kitchen window of their cabin. They were convinced they had seen a sasquatch and, when they found a tuft of coarse, dark hair very close to a large footprint seventeen inches long and five inches wide, they sent the hair to the Government of Yukon Department of Environment for identification. From its microscopic appearance the lab thought the hair was probably from an ungulate, maybe a bison, and passed it on to the University of Alberta for DNA testing. This confirmed the visual assessment that the hair was indeed from a large ungulate and probably a bison, and concluded that this was what had fooled the good people of Teslin.

As with the Milinkovitch *mehti* paper, I emailed the principal author, Dr David Coltman from the University of Alberta, to ask whether the article was genuine. My enquiry produced more of the back story and Dr Coltman told me how he had become involved:

> My attention was attracted because the regional biologist from the Yukon, Philip Merchant, stated that the hair closely resembled bison. I also happened to know Philip quite well and he is a very competent and diligent professional. But the question about whether the hair was really bison or not remained, so I called Philip and asked him to send it me and we would DNA test it so we could bring something definitive back to the press, but really for a bit of fun, but also because I saw this as a good opportunity to do some good outreach about science and wildlife genetics.

When the sample arrived, Dr Coltman successfully extracted and sequenced the mDNA, in fact the same control region that we have already encountered, and detected a clear match with bison. The story made headline news with a campus press conference, interviews on CNN and coverage by all the major TV networks. Soon afterwards he was having dinner with the editor of the influential *Trends in Ecology and Evolution* and she persuaded him to write a short article. Rather like Prof. Milinkovitch's paper on the Matthiessen *mehti*, Dr Coltman could not resist a titillating final paragraph:

> There are several possible explanations for these results. First, as suggested from molecular analysis of hair from a suspected Yeti (*referring here to the Milinkovitch paper*), the sasquatch might be a highly elusive ungulate that exhibits surprising morphological convergence with primates. Alternately, the hair might have originated from a real bison and be unrelated to the sasquatch. Parsimony would favor the second interpretation, in which case, the identity and taxonomy of this enigmatic and elusive creature remains a mystery.

Like all the best yeti and sasquatch tales, it ends with an element of uncertainty. Normally this ambiguity is genuine – indeed the extent of the uncertainty is usually played down. In this case, though, there was very good DNA proof that the hair came from a bison. Had there not been this lingering element of mystery, I doubt the press would have been as interested as they were.

There was one other thing that Dr Coltman told me. The witnesses from Teslin were certainly well aware what a bison looks like, even from behind when only its two hind legs would be visible. They had recently shot one, their freezer was full of bison steaks and the pelt was hanging up in their cabin. Was the temptation to start a sasquatch rumour too great? If so, then it worked. Not many CNN news crews travel to the town of Teslin, Yukon during the course of a normal year.

The one exception to these rather casual liaisons between science and cryptozoology is the Sasquatch Genome Project run by Dr Melba Ketchum. Dr Ketchum is a qualified veterinarian who operates a DNA testing company in Timpson, Texas. The Sasquatch Genome Project website reports that the project has been under way for several years. Over the last couple there have been periodical releases of the project's progress, though nothing was formally published.

These releases were always eagerly anticipated by the crypto-zoology community and when I met up with Bigfoot enthusiasts on my first tour of the US in early 2012, Dr Ketchum and the Sasquatch Genome Project was the one topic that dominated all others. Wherever I went, everyone I met wanted to know what I thought about it. I could only reply that until the research was published and I could have a good look at the data there was very little that I could say. That rarely put a stop to enquiries and was immediately followed by, 'Well yes, but what do you think anyway?' I did however admit that it is highly unusual for a professional scientist to release preliminary information about the outcome of a project intended for publication, except in confi-dence and for a good reason. Good quality science journals are reluctant to publish anything that is already in the public domain, and additionally, if there are intellectual property issues, disclo-sure will invalidate any patent application. So the early release of the Sasquatch Genome Project conclusions was highly unusual.

For example, the following explosive press release appeared in November 2012:

FOR IMMEDIATE RELEASE 'BIGFOOT' DNA SEQUENCED IN UPCOMING GENETICS STUDY
Five-Year Genome Study Yields Evidence of Homo sapiens/Unknown Hominin Hybrid Species in North America

DALLAS, Nov. 24 – A team of scientists can verify that their 5-year long DNA study, currently under peer-review, confirms the existence of a novel hominin hybrid species, commonly called 'Bigfoot' or 'Sasquatch', living in North America. Researchers' extensive DNA sequencing suggests that the legendary Sasquatch is a human relative that arose approximately 15,000 years ago as a hybrid cross of modern *Homo sapiens* with an unknown primate species.

The study was conducted by a team of experts in genetics, forensics, imaging and pathology, led by Dr Melba S. Ketchum of Nacogdoches, TX. In response to recent interest in the study, Dr Ketchum can confirm that her team has sequenced 3 complete Sasquatch nuclear genomes and determined the species is a human hybrid:

'Our study has sequenced 20 whole mitochondrial genomes and utilized next generation sequencing to obtain 3 whole nuclear genomes from purported Sasquatch samples. The genome sequencing shows that Sasquatch mtDNA (*mitochondrial DNA*) is identical to modern Homo sapiens, but Sasquatch nuDNA (*nuclear DNA*) is a novel, unknown hominin related to *Homo sapiens* and other primate species. Our data indicate that the North American Sasquatch is a hybrid species, the result of males of an unknown hominin species crossing with female *Homo sapiens*.

'Hominins are members of the taxonomic grouping Hominini, which includes all members of the genus Homo. Genetic testing has already ruled out *Homo neanderthalensis* and the Denisova hominin as contributors to Sasquatch mtDNA or nuDNA. The male progenitor that contributed the unknown sequence to this hybrid is unique as its DNA is more distantly removed from humans than other recently discovered hominins like the Denisovan individual,' explains Ketchum in the press release.

'Sasquatch nuclear DNA is incredibly novel and not at all what we had expected. While it has human nuclear DNA within its genome, there are also distinctly non-human, non-archaic hominin, and non-ape sequences. We describe it as a mosaic of human and novel non-human sequence. Further study is needed and is ongoing to better characterize and understand Sasquatch nuclear DNA.'

Ketchum is a veterinarian whose professional experience includes 27 years of research in genetics, including forensics. Early in her career she also practiced veterinary medicine, and she has previously been published as a participant in mapping the equine genome. She began testing the DNA of purported Sasquatch hair samples 5 years ago.

Ketchum calls on public officials and law enforcement to immediately recognise the Sasquatch as an indigenous people:

'Genetically, the Sasquatch are a human hybrid with unambiguously modern human maternal ancestry. Government at all levels must recognise them as an indigenous people and immediately protect their human and Constitutional rights against those who would see in their physical and cultural differences a "license" to hunt, trap, or kill them.'

Full details of the study will be presented in the near future when the study manuscript publishes.

If, as is suggested by the last line, the study was about to be published then any journal would be furious with this premature release of the content and would probably pull the article. Journals are naturally keen to make the most of the publicity generated by original articles, especially one as spectacular as this one promised to be.

In February 2013, the promised article was published, but not in a peer-reviewed scientific journal. It appeared in a new web-based publication called 'De Novo' and was the only article in the first and so far only volume.

I have been asked on numerous occasions to comment on Dr Ketchum's work, so I decided the best way to do it was to write as if I had myself been asked to review the 'De Novo' manuscript for a scientific journal. This is something I have done on very many occasions and I soon found myself able to go into full reviewer mode. The following paragraphs are extracts from my review of the manuscript:

To the Editor.

Thank you for asking me to review this manuscript. I must first declare a conflict of interest in that results from my own laboratory completely disagree with those set out in this MS.

The manuscript describes a series of DNA tests performed on various tissue samples attributed to the North American sasquatch, a large, bipedal and perhaps mythical creature reported to be living in America, especially in the Pacific Northwest.

The main, and dramatic, conclusion is spelled out in the abstract: namely that 'the data indicates (*sic*) that the Sasquatch has human mitochondrial DNA but possesses nuclear DNA that is a structural mosaic consisting of human and novel non-human DNA.'

Table 1 lists brief details of the 111 samples donated to the project. Three are saliva samples from either a food trap (#22,23) or a game camera (#27), one is mucus found on tree bark (#98), three are samples of tree bark (#47, 48, 49), three are dried blood samples (#25, 28, 107,110) and one is a toenail (#26). The other samples are hair, one with tissue attached (#18). This is a large sample and the authors are to be congratulated for assembling so much material.

There is a brief summary of the microscopic and electron-microscopic evaluation of the hair samples which claims that most of the hairs can not be attributed to any known species held in their 'reference collection'. This sweeping claim disregards the great difficulty in identifying the species origin of hair from microscopic appearance in most taxa, especially if the sample consists of one or a few hairs.

The authors state that the only hairs that were taken through to DNA analysis were those having visibly attached follicles. I am surprised so many of the donated samples passed this test as most were shed hairs that normally do not retain intact follicles.

The next stage was the extraction of DNA and sequencing of two regions of mitochondrial DNA, namely cytochrome b and hypervariable region I (HVS 1) of the control region. All 111 samples gave results at both cytochrome b and HVS 1 segments, which is an astonishing achievement. All samples revealed a human cytochrome b sequence and one of a range of different human sequences at HVS 1.

From these results and some unclear nuclear DNA data the authors conclude that the sasquatch is a hybrid between human females, which, in the authors' opinion, accounts for the presence of maternally inherited human mitochondrial DNA, and males of another unidentified species.

The real problem with this paper is that the authors have interpreted all the DNA results as supporting the hybrid genome hypothesis, while disregarding any alternatives. For example, the other parental species of the hypothesised hybrid is never identified and only assumed to be present at all because the nuclear DNA sequencing is not as one might expect from a good human sample. In my opinion the much more likely reason is that these are highly degraded and contaminated samples, despite the authors' efforts to argue otherwise. Such material is notoriously difficult to sequence and the authors have provided no convincing proof that their results are not entirely created by these well-known difficulties. In comparison, the efforts to by-pass human and environmental contamination in the sequencing of the Neanderthal genome (Green et al. 2010) took years of work to achieve.

The mitochondrial HVS 1 results show a wide range of sequence types whose continental origin can be deduced from their known geographical range. Of the twenty-five reported, nineteen are typically European, two are African and four are either Asian or Native American. To interpret these results as proof that the hybrids crossed the Bering land bridge from Siberia into North America is absurd. Despite the reassurances of the authors that they had eliminated all human contamination, it is far more likely that the majority if not all of the mitochondrial DNA sequences reported are the result of just that – human contamination.

This MS is not well written, and the project planning is poor. A lot of the material in this MS is unnecessary. It is a manuscript about DNA and any new version should restrict itself to exactly that. We do not need images of hair, trees, shelters, indistinct video clips and that sort of thing.

My advice to you is to reject this MS without offering the opportunity for revision with the same data. While the topic is certainly of great interest, the data do not support what is theoretically a most unlikely hybrid origin for Sasquatch.

My advice to the authors is to concentrate on the best sample and obtain a full genome sequence with at least 20x coverage. From that, and especially from any contiguous segments containing both genomic contributors, identify the other parental species, if indeed there is one.

Yours sincerely

Bryan Sykes

The Ketchum study never gets close to providing the exceptional proofs that such exceptional claims require. It may have begun as a promising project, but it was very poorly executed, has wasted a lot of valuable material, caused a great deal of confusion among cryptozoologists and must have cost someone a lot of money.

Caveat donor.

15

The Hunt Begins

Having covered the very patchy impact of genetics on the field of cryptozoology, let me at last describe how Michel Sartori and I set about making our own contribution. Soon after agreeing the collaboration between Oxford and Lausanne we began to discuss the practical issue of getting hold of sufficient material to make the project worthwhile. I had thought there might be some useful samples in the Heuvelmans archive itself, but there was not. Michel and I agreed that obtaining enough material might be a serious obstacle to the project's success. We had to come up with a good plan. We decided in the end to issue a joint press release, announcing the collaboration and inviting individuals and institutions to submit samples for testing. Although we settled on this approach in the summer of 2011, it was not until the following spring that we were both sufficiently free of other commitments to cope with a large response – if there were to be one.

We drafted the following announcement:

PRESS RELEASE

SCIENTISTS SEARCH FOR YETI DNA
The Oxford-Lausanne Collateral Hominid Project

<u>Background</u>

Ever since Eric Shipton's 1951 Everest expedition returned with photographs of giant footprints in the snow there has been speculation that the Himalayas may be home to large creatures 'unknown to science'. Since then, there have been many eyewitness reports of such creatures from several remote regions of the world. They are variously known as the 'yeti' or '*migoi*' in the Himalayas, 'Bigfoot' or 'sasquatch' in America, '*almasty*' in the Caucasus mountains and '*orang-pendek*' in Sumatra, as well as others. Theories as to their species identification vary from surviving collateral hominid species, such as *Homo neanderthalensis* or *Homo floresiensis*, to large primates like *Gigantopithecus* widely thought to be extinct, to as yet unstudied primate species or local subspecies of black and brown bears.

Mainstream science remains unconvinced by these reports both through lack of testable evidence and the scope for fraudulent claims. However, recent advances in the techniques of genetic analysis of organic remains provide a mechanism for genus and species identification that is both unbiased, unambiguous and impervious to falsification.

These techniques were not available to biologists like Dr Bernard Heuvelmans, whose 1955 book *Sur la Piste des Betes Ignorees* (translated into English as *On the Track of Unknown Animals*) helped foster widespread public interest in the subject. Between 1950 and 2001, the year of his death, Dr Heuvelmans, as well as investigating numerous claims, assembled a considerable archive that is now curated by the Museum of Zoology in Lausanne, Switzerland.

In this release we are pleased to announce the launch of the Oxford-Lausanne Collateral Hominid Project, a collaboration between the University of Oxford and the Lausanne Museum of Zoology to employ these new genetic techniques systematically to investigate organic remains from these and other cryptozoological* samples. We invite submissions of material, particularly hair shafts, for analysis accompanied by details of their provenance. For submission procedures please visit *http://www.wolfson.ox.ac.uk/academic/GBFs-v/OLCHP.*

The principal investigators are Bryan Sykes, Professor of Human Genetics at the University of Oxford and Michel Sartori, Director of the Museum of Zoology, Lausanne.

Cryptozoology: The search for animals whose existence is not proven.

The press release was sent out on 22 May 2012. Nothing happened until two days later, when I was contacted by Reuters, the press agency. After their report went out on the wires, the floodgates opened. Both Michel and I were inundated by interview requests from newspapers, radio and television.

I was pleased to see that there were very few comments about whether or not this was an appropriate subject for investigation by 'serious' scientists. The only one I encountered came from a professor of anthropology in St Louis, Missouri, and which was put to me in an Associated Press interview. He considered any research in this area to be both frivolous and futile but I responded that he had basically misunderstood the philosophy of science. I was making no claims that anomalous primates existed, or did not exist, but doing what science is all about – finding and testing evidence. I added that I did not have to believe what I was told, or even form an opinion. I just needed to test the evidence. Michel had a similar response from a professor of paleogenetics from Mainz, Germany who said in a radio interview: 'I only fear that the examined questions, which are not really relevant to zoology, but rather belong to the boulevard press and can present some interest to the public, will never be discussed by serious zoologists.' Other than those two rather mild admonishments, I was pleased to see that we had not stirred up much resistance to our project.

This media interest rumbled on during the following months. Michel and I were filmed in Lausanne by NBC for the *Today* show, which went out to their audience of 100 million people all over the world. The magnitude of the response at least showed that what we were doing was popular. Perhaps even more important, it also brought a good number of responses from individuals who told us about material they had that they were willing to submit for testing. In no time, my office in Oxford and Michel's in Lausanne were filling up with all sorts of yeti samples. We knew we would have enough material for a decent project.

However, not all of the samples were submitted in a formal way. I had specified that potential donors should email some details of what material they had, then Michel and I would make a decision about whether to include it in the study. Most contributors complied admirably and returned their hair samples in one of the sample bags that I sent out to them as a package, complete with a pair of disposable gloves. This was designed to prevent contamination through handling; although I could remove that, I hoped it would also give contributors the impression that they should be careful. Each proper submission was accompanied by a numbered consent form that gave us permission to analyse the sample, and confirmed that the donor was in a position to donate the sample to the project. Michel had kindly agreed that, when the project was over, unused material could be accessioned in the Heuvelmans archive in his museum. Contributors could choose between having their material kept in Lausanne for further study in the future, or returned to them after analysis. The great majority chose to have their unused material accessioned. As I was to discover as the project proceeded, so many samples that I heard or read about had been lost or could not be located. Michel and I were keen to make sure this didn't happen to our samples and that they would be made available to researchers in the future.

All these precautions did not prevent the arrival of samples without warning and unaccompanied by a consent form or any other documentation. One package arrived in Oxford from Colorado containing three hair samples, each in Ziploc bags, which was fine. But they were not labelled, there was no indication which sample was which, nor any information about where they had been found or what made the donor think they had come from a Bigfoot. There was a note with a name and address, but nothing else, so at least I was able to return them to the owner with a request that he label them, give me some information and send them back. They never arrived.

Although I had agreed a good rate for the DNA analysis at Terry Melton's lab, it was not so low that I could send every sample off for testing without a second thought. At least I thought I should have a look at them under a microscope. It was a good job I did, because they were obviously not all hairs. One looked like plant material, roots of some kind, with clearly branching fibres which is not a feature of mammalian hair. Another sample looked very odd indeed. The fibres were dead straight, very thin with a satin sheen and no visible internal structure. I set aside the samples I wasn't sure about and took them to the US Fish and Wildlife Forensic Laboratory in Oregon later that year before selecting what to send to Terry's lab for DNA analysis. This kind of visual screening certainly saved the expense of pointless testing, but if I was in any doubt and if the provenance was good, off it went to the lab in any case.

As time went on, and news of our project spread through the Internet, samples arrived in Oxford at regular intervals. Even though I had made it clear that I was only going to work on hairs, this preference didn't stop a range of other materials being offered, or in some cases arriving in Oxford unannounced.

More than once I was asked if I would be able to analyse objects that donors believed had been touched by a Bigfoot. One I recall was from a lady in Tennessee with a family of Bigfoot living on her farm. This family had been there for many years and she knew them all by sight, though oddly had never managed to take a photograph. She used to leave sausages for them on the back porch at night in a plastic sandwich box. In the morning the sausages were gone, and there was saliva and bite marks on the sandwich box which she kindly offered to send me for DNA testing. I declined, and not just because this sounded like a very tall story, but because I could not decontaminate slime on a sandwich box in the same way I could a hair.

I did, however, accept one saliva sample and that was because the story behind it was so striking. This was from a gentleman in Washington State who had been driving home one night when a Bigfoot leapt onto the bonnet of his car and tried to attack him through the windscreen. In so doing the creature left a layer of slime on the screen before it fell off. The driver, realising he may have proof of Bigfoot, had the windscreen removed and stored it in his garage wrapped in cling film. I was so enthralled by this story that I sent my correspondent a DNA swab with instructions to get a sample of slime from the windscreen. This was duly returned and sent off to the lab. A week later the result came back with a definitive identification of the creature. It was a cow.

Another case of an unorthodox sample came to my attention when my contacts in the Bigfoot world alerted me to the excitement surrounding a sasquatch skull that had been discovered on the Colville tribal area of eastern Washington State. This was obviously an extremely important find as it would not only allow a DNA identification but also a proper anatomical investigation. Was this the vital piece of evidence that the world had demanded for so long? If so, when the sceptics asked 'Show me a body,' the Colville skull could be produced with a flourish. I was getting excited by the prospect, and so I asked my contact to send me a photograph of the skull.

It duly arrived, not crystal clear but in good enough focus to see the top of the skull. The feature which had immediately convinced my contacts of its Bigfoot origins was the prominent sagittal crest, a bony extension which ran across the top of the skull. This is a feature of all great apes, especially the gorilla, where it secures one end of the massive jaw muscles the great ape needs to chew its food. At a stroke, so it seemed, the mystery of Bigfoot had been solved. It was a great ape, like a gorilla and, since its sagittal crest was even more pronounced than that, this Bigfoot was most likely a vegetarian endowed with the muscular equipment to pulverise the toughest plant.

The Colville skull was lacking any facial bones or teeth, but I was not unduly disturbed by this omission. Who knows how old it may have been and, in any event, many skulls are fragmented when they are first discovered. For example the Neanderthal skull unearthed in Spy in Belgium in 1886 had the top of the skull separated from the facial bones. The frontal bone of the skull of a Neanderthal child from La Cariguela de Pinar in southern Spain is similarly detached from the face. I was about to impress upon my informants that this skull must be secured for science at all costs, and even considered offering to help finance an expedition to Colville to recover this unique piece of evidence, when I had second thoughts.

It was Christmas, and as usual we were having turkey. On Boxing Day the remains of the bird went into the pot to make stock. After it had simmered for most of the day, the now de-fleshed carcass was lifted out of the saucepan and put on one side ready for disposal. I must have been thinking about the Colville skull because, looking at the carcass, I could see the same sagittal crest on the Christmas turkey as I had seen on the Bigfoot specimen. I rushed through to my computer, and looked again at the image of the Colville skull. It was similar, but not identical to the turkey. A little relieved, but not much, I hesitated in offering to cover the cost of the expedition to recover the Colville skull until I had a second opinion. If it wasn't a turkey, then was it perhaps a goose? There was only one way to find out, so I bought one, cooked it and cut the flesh away from the breast. And there was the Colville skull. There was no doubt about it. The sagittal crest certainly supported muscles, the flight muscles of a goose rather than the powerful jaws of Bigfoot. The odd indentations around the Colville skullcap were the insertion points for the ribs. I emailed my contacts to call off the expedition, only to hear that Dr Jeff Meldrum had also come to the same conclusion as to the skull's anserine origins.

It was around this time that I agreed terms for a television documentary with Harry Marshall of Icon Films that would follow the yeti project. It had been Harry who, ten years earlier, had made the film about the Bhutanese *migoi* I mentioned earlier. It had been beautifully made and I had enjoyed working with him. However, the new documentary was an enterprise on a quite different scale. Whereas my contribution to the Bhutan film had been limited to the three DNA analyses carried out by my lab, this time Harry persuaded the UK broadcaster Channel 4 to commission three one hour films covering the entire project. That was very good news of course, but it did change the time-table of my own research. Naturally, Harry wanted to know the results of the DNA tests I was carrying out on the specimens that I had already received. I was keen *not* to know the results until after I had spoken at greater length to the sample donors. I was sure that had I known the test results before I interviewed the donors, it would influence the direction of our conversation. Also, I needed to get out into the field on my own, unaccompanied by the film crew, and try to dig up some more interesting material. I had always planned to do this, but the filming schedule for the documentary made me act more quickly than I otherwise would.

I am very glad I did so, as you will gather from the following chapters, as I introduce you to the characters I met on my several journeys. I never travelled without my voice recorder so most of what you read is taken from recordings with only the lightest editing.

16

The Guru

The majority of hair samples donated to the collateral hominid project came from North America, and I made sure I visited and interviewed most of the donors. I also spoke to a lot of people who had encountered Bigfoot in one way or another, either through a sighting or an 'experience' such as you will read about in the coming chapters. The Bigfoot enthusiasts were a fascinating bunch, all keen to help with collecting the evidence to identify the creatures that they never doubted for a moment were living in the forests and mountains. As I travelled around I became absorbed in their community, no longer debating whether or not these creatures existed but rather learning about the finer points of their social structure, even their language and culture. As time went on I, like the enthusiasts, began to take it for granted that we were discussing a real creature, only very occasionally pausing to remind myself that there was no proof. The time and energy that the enthusiasts put into their research was impressive. Days or weeks spent in isolation deep in the forest staking out a Bigfoot hotspot, hours on the Internet catching up with the latest sightings, or just to gossip. They were the nicest bunch, charming and open, but were they all deluded? I will let you judge for yourself.

Many people have spent years involved with Bigfoot research, but none for quite as long as Loren Coleman. He studied zoology and anthropology at Southern Illinois University and Brandeis and had lectured at universities in New England until 2004. He now lives in Portland, Maine where he established the International Cryptozoology Museum in 2003. I had two good reasons for wanting to meet and interview Loren. First, he had written a thoroughly researched biography of Tom Slick Jr, the wealthy Texas businessman who financed so many of the yeti and Bigfoot expeditions in the 1950s and 1960s. Coleman calls him 'The Howard Hughes of Cryptozoology'. And second, with a lifetime spent in and around cryptozoology, I wanted his assessment of the current state of affairs.

I had arranged to meet him in Portland in April 2013, travelling up from Boston where I was speaking at the annual meeting of the New England Historical and Genealogy Society about my recently published book on the genetic history of America, *DNA USA*. On Monday 15 April two bombs exploded near the end of the course of the Boston marathon. This awful event, which killed three people and injured a further 264, happened in the week of the genealogy symposium. My wife and I arrived on Thursday 18 April and booked into the Four Seasons hotel on Boston Common where, despite the bombing, the genealogy event was scheduled to take place. Almost as soon as we checked in, news came that two local residents of Chechen origin, the Tsarnaev brothers, had shot and killed a policeman in Watertown, a suburb of Boston. One of the brothers was dead, but the other had escaped. The manhunt for the suspected bombers started immediately.

Suddenly central Boston emptied. The normally busy streets around the hotel cleared of traffic and everyone was glued to their televisions. Like most people, I expected the second suspect to be found and shot within hours, but into the night no news came. The next morning, when I was planning to take the train to Portland to visit Loren Coleman, a note on yellow paper slipped under our bedroom door informed us that the governor had issued an unprecedented 'shelter-in-place' order. Boston was in lock-down while thousands of police searched the streets and homes. No one was allowed out of the hotel.

It was a strange feeling; I could sense the whole city coming together to hunt down the bomber. All day we continued to expect to hear at any moment that he had been found and killed. Of course, if you are going to be in lock-down, the Four Seasons is not the worst place to be. However, at the impromptu lunch which the hotel had organised with the skeleton staff that had been able to get to work, we sat a long way from the windows overlooking the park, just in case of another shoot-out. At 7 p.m. the 'shelter-in-place' order was rescinded and shortly afterwards the second bomber was found and arrested.

Probably the least important consequence of the manhunt was that I was unable to get to Portland to meet and interview Loren Coleman. I had to leave for other Bigfoot appointments on the West Coast. Fortunately my researcher Marcus Morris was able to catch up with Coleman in Salt Lake City and put to him the questions I would have asked, had it not been for the lock-down in Boston.

I was aware that Coleman had a rare sample from the 1959 Slick expedition in his collection and had asked him if I might test it. He agreed and he sent me three hairs – very precious specimens indeed, and with a suitably cryptic description. The hairs were labelled 'Animal X'. Coleman explained that the Slick expeditions had always referred to their quarry not as a yeti but as 'Animal X' and this was reflected in the labelling. As the Slick records vanished after his death, Coleman is not too sure about the provenance of this particular sample. It came from the late George Agogino, the scientist appointed by Slick to distribute the expedition samples among his panel of experts.

Agogino had clearly examined this 'Animal X' hair sample under the microscope, but according to Coleman had been unable to identify it. Agogino had, however, been quick to recognise several of the hair samples brought back by Sir Edmund Hillary's 1960 yeti expedition to the Himalayas that we covered in an earlier chapter. The hairs he examined from the Hillary expedition, Agogino concluded, belonged to *Pseudosis nayaur*, the Himalayan blue sheep, a major prey item of the snow leopard and the focus of Peter Matthiessen's 1992 expedition that brought back the *mehti* hairs. But 'Animal X' was not a blue sheep. More than fifty years after the hair sample was brought back from Nepal, the identity of 'Animal X' was still a mystery. With only three hairs to work on it was a tall order to get enough DNA for a species identification, but given the enduring mystery and the historical relevance of the sample, it was definitely worth a try. And, as you will hear later, it worked.

I hope Loren Coleman and I meet before long, but in the meantime I wanted to have his thoughts on the current state of the search for the yeti and Bigfoot. As well as being an active researcher himself, his long experience and his wide-ranging interest made him someone whose opinion I wanted to have. As he told Marcus Morris:

'This is a very exciting time. There are a lot of new people in the field, there's lots of controversy of course and a lot of polarization. Are you in the kill or not kill faction? In North America right now there's a raging debate. If you find a Bigfoot do you kill it or do you just study it? The capture and release approach really has more and more basis in fact. There's large groups of individuals who won't even join an organisation if they're pro-kill.

'On the other hand the respected Washington State University anthropologist, the late Grover Krantz (*he died in 2002*) came in for a lot of criticism because he strongly advocated killing a Bigfoot in order to establish it as a new species. He even built his own helicopter to cruise above the forests of the Pacific Northwest so he could shoot one.

'That's all well and good, but we see nowadays that sometimes we may be shooting almost the end of the species. So the new approach is, let's get close to one, let's capture one, let's test it for DNA, take a lot of photographs, maybe even put it in a reserve. And so those older line individuals with their guns and with the killing are really moving to the back end of the debate. There are younger kids coming forward, more women are coming into the field, more nationalities, and they're bringing a more global approach of not killing these creatures, just studying them.'

Then there are the sharp divisions of opinion about whether Bigfoot is a *Gigantopithecus* or a Neanderthal or some other hominid. There are also those who have an altogether more spiritual approach to the whole field.

'I call some of these individuals 'Bigfoot contactees'. They're very much in a religious state of being where they feel like they're having a conversation with Bigfoot. That they're in touch with Bigfoot, that they're actually reading the Bigfoot's mind or diagnosing, you know, what the Bigfoot's thinking and saying on a level that there's absolutely no evidence for. You know some people say they've had contact over decades with Bigfoot. And you ask them for a picture of the Bigfoot and they say, well, he wouldn't let me take it. To me that's talking about religion, and true belief and cult, not about science.'

When I heard this I was reminded of a conversation I had with Rhett Mullis, my guide and companion in the Pacific Northwest. Rhett's organisation, Bigfootology, collates reports of Bigfoot sightings and interviews the witnesses. On one occasion, a lady reported seeing an old Bigfoot walk right into the ocean on the coast of Oregon. She described at length how she and the Bigfoot walked slowly across the sandy beach and into the water. She stayed on the beach but the creature waded further and further out until it disappeared beneath the waves, clearly intent on ending its own life. Rhett's witness took at least forty minutes on the phone describing this moving scene in great detail.

Understandably, Rhett was very excited by the prospect of finding, at the very least, some fresh footprints in the sand above the tide-line and maybe some strands of hair snagged on a bush where the Bigfoot had broken cover. So he asked his witness where exactly this beach was and if she could take him there. 'Well,' she replied, 'I didn't actually see it in the technical sense. I was there all right, with the Bigfoot, but it was more of a kinda mind-melding experience, you know.'

Rhett politely ended the conversation, put the phone down and let out an enormous groan.

On my travels meeting Bigfoot enthusiasts I have come across my fair share of what Coleman would call 'contactees' who have turned evasion and circularity into a fine art. Often this shows in answer to a direct question like:

'Why do you think no one had ever found a body?'

'Well they bury their dead underground, don't they?'

'How do you know that?'

'Because you never find a body.'

The same line of logic surfaces when the topic turns to the widespread failure of infrared-activated trail-cams to capture an image.

'They are very intelligent. They can detect infrared and learn to avoid it. They can also sense human intent, you know.'

'How do you know that?'

'Because you never get an image on a trail-cam.'

I had a very similar experience to Rhett when a charming couple came to see me in Oxford. The husband was a former nuclear security guard in the US Air Force and a lifelong Bigfoot enthusiast. His wife, in common with most 'Bigfoot widows', was tolerant at best. I was particularly keen to talk to this gentleman who, as a boy in 1968, had seen the Minnesota Iceman in Frank Hansen's travelling show in the Midwest.

I listened intently as he described seeing the Iceman frozen in a block of ice. He certainly thought it was genuine and very scary, and it was this experience that had triggered the fascination with Bigfoot which had never left him. As our conversation was drawing to a close he suddenly started to tell me about the Bigfoot that used to visit Edwards Air Force base in California when he was still working in nuclear security. 'That sounds odd,' I thought to myself. Even I had heard of this base and its fearsome array of military aircraft ready to fly around the world at a moment's notice and unleash a nuclear holocaust. How come the Bigfoot were able to penetrate what was presumably one of the most secure places on the planet?

'They live in the tunnels under the base,' was the reply when I put this puzzle to my guest.

'Don't they get caught on security cameras?' I asked.

'No, they can sense the cameras and flip into another universe. They have a shape-shifting capability,' he replied.

I refrained from asking how he knew they were able to do this. I was certain to get the reply that it was because they were never seen on the cameras.

Back to Loren Coleman. I wanted to know about the human cost of being a Bigfoot enthusiast, especially if your partner is not.

'Definitely people have lost their wives, lost their jobs, lost their husbands. Some people are fired, they get very upset, and yet they want to hang in there. They have to have big egos. So there's a whole list of things that really calls for an individual to have a thick skin. To survive in this field it takes an individual that's very strong-willed and clear on what they want to be doing in it. So there is definitely a dark side to Bigfootology.

'There's also people in this who seem to be hoaxers and money grabbers and drifters and all kinds of people. The media love this, of course. Their approach is that all of the Bigfoot field is full of hoaxers. But that's really wrong. Research, field studies are really not that dominated by hoaxers. It's only one percent, but that one percent gets all the publicity and really ruins it for the rest of us. It's the hoaxes that get all of the attention. It makes for great television.'

Avidum genus auricularum.

17

The Mountaineer

I only ever got one prize at school. It was for chemistry in the Lower Sixth form. We could choose a book and, after a lot of searching, I asked for *The Encyclopaedia of Mountains*. It is still here on my bookshelf and every so often I take it down and read a section on some far-off peak. Of course, it is very out-of-date now and the photographs are poor by modern standards. But it still brings back the thrill I felt at reading the stories of great adventures. Like the three escaped Italian prisoners of war who, in 1943, made it to within 500 feet of the summit of Mount Kenya using only the picture from a tin of Oxo as their guide. Or the heroic first ascent of the notoriously difficult Nanga Parbat in the eastern Himalayas by the Austrian Hermann Buhl climbing alone and without oxygen in 1953. This was all schoolboy fantasy, and my own climbing career never exceeded Monte Rosa in the Swiss Alps. I considered attempting the nearby Matterhorn, but all I saw was death. So you can under-stand my excitement when I was on the way to meet Reinhold

Messner, without doubt the greatest living climber – some say, the greatest climber in history.

Messner, as many of you will know, was the first person to climb Everest solo and without oxygen, but that is only the most celebrated of his mountaineering achievements. He was, for example, the first to climb all fourteen 8,000 metre summits and followed Hermann Buhl's alpine style of small climbing parties, even in the Himalayas, rather than joining large-scale expeditions that 'laid siege' to a mountain until it submitted. Sadly, he lost his younger brother Gunther on the descent from Nanga Parbat in 1970. He also lost seven toes to frostbite, but that did not stop him climbing. His numerous achievements since are the stuff of legend, but they are not what brought the two of us together.

Among the many books Messner has written about his adventures is one entitled *My Quest for the Yeti*. In it he describes an experience in the summer of 1986 while trekking through eastern Tibet. At the start of the book he mentions that he has a pelt and a head from a yeti, and that he was making these available to scientists to study. I decided to see if he would let me test some of this material.

It is no easy matter getting an appointment with Reinhold Messner. I emailed his admirably protective secretary in the southern Tyrol near Bolzano where he lives, asking for a meeting. I mentioned that I was the scientist who first recovered DNA from Oetzi, the 5,000 year-old Iceman found in a glacier at the head of the Oetzal Valley near Messner's home. I knew that Messner was among the first on the scene and had, in fact, been the one to realise its great antiquity. Before his intervention, the body was believed to be that of an Italian music teacher who had got lost in the region in the 1930s. So there was at least a nominal connection between us. That didn't immediately open any doors and I eventually decided that the only way I was going to get to see Messner was to turn up on his doorstep.

Though he still climbs, Messner has been hard at work developing five museums in South Tyrol, each devoted to different aspects of the mountains and the people who live among them.

The principal museum, and Messner's HQ, is in Sigmundskron Castle perched high up on a rocky promontory above Bolzano. As soon as Ulla and I arrived at Bolzano one hot August day I rang Ruth, his secretary. Yes, Messner was there, but may not have time to see me. When we arrived at the castle gates we were told to wait in the open-air café in the grounds and Messner might be able to spare five minutes. I knew Messner had a reputation for irascibility; sure enough, when he rounded the castle wall and came into view he looked like thunder. His abundant hair framed a craggy face familiar from old photographs. The expression was far from friendly.

I began to explain why I was there and how I wanted to run DNA tests on his yeti samples. All the time I was talking, I was trying to concentrate on the man sitting opposite. Here I was with the greatest mountaineer in the world, the man who had climbed Everest on his own. It was hard to take it all in. I began by saying that I had read his book and that I had developed a new way of analysing hair that might identify the yeti specimens he had in his collection.

'It's a bear,' was his only reply. I could see he was completely sick of talking about yetis; in fact, I imagined he might have regretted having written a book about them at all. Nonetheless, I carried on with my explanation.

'It's a bear,' was once again his only response, in his strong German accent. Tyrol might technically be part of Italy these days, but its language, history and customs are much closer to Germany and Austria.

'Yes, probably, but what sort of bear is it?'

His impassive grey eyes flickered momentarily with what looked as though it may have been interest. I pressed on and began to get more assertive. He might be the world's greatest climber, but I was also high up in my own field and I had come a long way to see him. Looking back, I was rather glad he was so grumpy. It served to banish my nerves far more effectively than any politely condescending smile would have done. Eventually I asked Messner if I could take a sample from the yeti. If I left without one it would have been an experience, but I would have felt forever frustrated that I had failed.

He looked at me again. The cumulonimbus had cleared ever so slightly from his expression.

'Meet me at Juval at three o'clock.' Then he got up and walked back around the castle wall and out of sight.

So at three o'clock Ulla and I found ourselves being driven by taxi up an even steeper hill to an even rockier crag, on top of which was an even more precariously perched castle, Juval. This is Messner's family home during the summer months. He led us up the steep incline to the gates and I tried my best not to appear out of breath. He took us through the gates and into a room leading off from the courtyard. There on the wall, in the half-light, was the mounted head of a yeti.

I already knew its history. This was one of two yetis in Messner's collection that were brought back by the Nazi-inspired expedition to Tibet led by SS officer Ernst Schafer in 1938 and sponsored by none other than Reichsführer SS Heinrich Himmler. Himmler had many strange ideas, one of which being that the Aryan race was forged by a combination of ice and fire in Shambhala, a lost kingdom in the mountains of Tibet. The yeti, Himmler imagined, might be an ancient product of this fusion and so he sent Schafer off to find one. Schafer thought it was a pretty crazy idea, but he was a zoologist and, as today, any source of research funding is welcome. The expedition carried out a lot of what is now outdated anthropological work, including measuring the skull shapes and sizes of the local people. Schafer also shot a lot of yetis and mocked the Tibetans for being afraid of what he could clearly see were bears. When Schafer died in 1992, his widow, keen to get the mounted yeti head and other grizzly relics out of the house, offered them to Messner. Even though they were by now over seventy years old, I thought there was still a good chance of being able to recover some DNA from a hair sample. With Messner holding the evidence bag, I reached up and snipped a few hairs from the neck of the stuffed yeti head.

Messner's second yeti was at yet another museum, at Sulden,

an hour further up the valley, and it was closed at the time. But now we had got to know each other a little, arranging a return visit to take a sample from the second yeti was far more straight-forward. When I returned the following summer there was also more time to talk to Messner about the encounter that he himself had with a yeti back in 1986. First, I asked him to tell me more about what he was doing in Tibet.

'I was crossing Tibet from the east to Lhasa and then going further on to Nepal, following the trail of the Sherpas. The Sherpas left eastern Tibet 450 years ago and travelled for over seventy years before they settled in Solokhumbu in Nepal. We still don't know why they did it. A German archaeologist was following this Sherpa migration in his studies and I read about it and was interested. Also I had never been to the eastern part of Tibet. It was also closed to foreigners and I was not allowed to go there. But I sneaked in. I hid in a monastery and from there I started going west, going up and down in yak caravans, but also some-times with cars if it was possible.

'And once, late in the evening I was hoping to find a small village to have a night with the local people, also hoping to get some food. Then I could see a strange figure in the woods. It was beginning to be dark and I was hoping maybe in ten minutes I will find a place, but there was no place. This strange figure went away, not running away. It was in the last light of the day so I could not see immediately what it was. It was too quick. I could see it and then it was gone. I was remembering a shadow in my eyes and I asked myself have you really seen something or was it only imagination? And when I went to where the figure was standing I found the footprints. I knew there was something real, it was not only an imagination. They were exactly like the foot-prints in Shipton's photograph. And it was the first moment I was thinking this is very strange. It looks like the footprints of a yeti.

'But I was not yet thinking about a yeti seriously. I was afraid because it was a huge being, I couldn't even say it was an animal, human being or something else. And I became afraid and I went and didn't find any villages and I went about the woods, I was up maybe 4,500m, no village and I was so tired that I tried to give up and in this moment again in the moonlight I could see a similar figure on two legs, going away. Not attacking me.

'I was very frightened, and I became more and more frightened when I was trying to find a place to sleep, but I made a bivouac, staying outside. There was one river behind me so I could not go back, and one river in front of me that I had to cross to get to the next village. And this river was very wide coming from maybe some glacier mountains and there was no bridge. I couldn't cross it. For me crossing a river is not so easy because I lost part of my toes, so being barefoot in a river is difficult. It was too dangerous in the night and I went back and I tried to find shelter and sleep, a few rocks. And I could not sleep because otherwise this creature is coming and killing me or eating me, whatever, I didn't know what it was.

'Very early in the next morning and I crossed the river and found a village and it was empty. No people around, only a few dogs. And I went in the end to a house and went to the upper part and tried to sleep. But local people came with fires, and they told me to come down. They took everything I had away. I was really afraid that something might happen or they might kill me. I began to defend myself by saying that I was running away from a huge being. And then I heard the first time the name *chemo* (*Messner pronounces it as 'chay-mo'*) and they taught me, they understood my eyes looking there and movements that I was in touch with the *chemo* but I didn't know what is the *chemo*. And after this I was beginning to try to understand my possibilities, very poor possibilities to communicate from the local people what is this *chemo*, and I had a feeling maybe it's something like we call the yeti. Only afterwards I was speaking with the local people. I understand they have a great respect for this creature. They spoke with fear about it and also had some compassion with me because I was able to run away from this creature. So with this I made friends with them, otherwise maybe they would kill me because I came in the night in their villages without even asking them if I could sleep there. They pulled me in the houses and gave me food and asked me about the *chemo* I had seen and how big and where it was.

'They said they are very, very dangerous but normally hiding so you cannot see them. They appear and then disappear again. For me there is no doubt that the yeti legend is based on this special bear. And the local people are speaking about a huge, huge being, bigger than a human being. They speak about a hairy figure, a stinky figure going on four legs, two legs, leaving footprints like human.'

Like other witnesses, this experience had a profound influence on Messner and over the next few years he went back again and again to the Himalayas. To climb of course, but also to find out more about the *chemo*.

'I went back to Nepal one month later. I was very interested in knowing what happened in this sighting. I went to many, maybe a dozen local Tibetans who are now living in Kathmandu in Nepal. And I knew them and went in their houses and asked them all the same questions. How are you calling in your homeland, in eastern Tibet, what we call, and what the tourists are calling yeti, in Nepal? And they all answered immediately "*chemo*" and so I know I have seen the animal which is the basis of the yeti legend.'

We returned to the subject of the footprints that Messner saw shortly after his experience. His first thought was that they were very like the footprint in the Shipton-Ward photograph and definitely not made by a bear. Later on, as he went in search of the *chemo*, he began to form an idea that this animal actually was a bear, albeit a special kind of bear. He had taken a photograph of the print he found in the forest and compared it to the Shipton print.

'With a lens it was possible also to see the nails of the fingers. I could see it was exactly like the Shipton footprint. The Shipton footprint is in the snow and the snow around the Shipton footprint is melted out badly by the sun so you don't see perfectly the fingers and nails of the bear. When I came home and I understood that the *chemo* is the yeti I could see this (*the Shipton print*) could perfectly be a footprint left by a *chemo*, by a bear.'

Messner was talking about a double print, where the hind foot is superimposed on the impression of the front. But, from his own experience, he came up with a very plausible explanation.

'I could see it later in the next years that this *chemo* is always walking like this when it comes on snow, crossing glaciers. He's putting the back foot in the fore foot because he knows instinctively that if the fore foot is standing (*here Messner put his right hand on the table*) so there's no crevasse, so he can safely put the other foot in the same place. So there's only footprints looking like a two-legged figure.'

I could now see why, at our first meeting, his only words had been: 'It's a bear.' Messner had been back to the Himalayas several times and asked about the *chemo* and is now completely convinced that it is some kind of bear. Whereas I, being a scientist, was curious to know what sort of bear it was, Messner was far more interested in the mythology surrounding this creature than its precise species identification, as he explained.

'I think for understanding the yeti story scientifically it is not so necessary to know the genes of the *chemo*. It's not important, the biological and genetic facts, it is important to study how legends are beginning. The legend is existing in the mind of the local people, but the legend always has a real base. Not an invention; all legends have a real base. The legends we have still in our memories which we heard from the grandmother and from the grandfather. And in this case after a while I understood that this is the way to understand the story and it's a different kind of bear.'

To Messner, the yeti and the *chemo* are one and the same. A legendary creature, but with a physical presence of a kind of bear. Actually I found Messner's zoological knowledge of Himalayan bears rather sketchy. In his book he shows two photographs of a *chemo* in captivity in Norbulingka Zoo in Lhasa, the Tibetan capital. One clearly has the characteristic white markings on its chest of the Asiatic black bear *Ursus thibetanus*. But this pale marking is also found in the lower altitude sloth bear, *Melursus ursinus*. The sloth bear has a longer coat and a bigger head than the black bear and the second photograph looks more like a sloth bear to me. Whether it is or not doesn't matter a bit, but what I did initially find quite surprising is that Messner didn't know what a sloth bear was. But when I understood that he was far more interested in how the yeti legend had spread and how it still has an influence on the everyday lives of the people he met on his travels, his skimpy knowledge of the details of ursine taxonomy seemed entirely unimportant.

Nevertheless, Messner did allow me to go to his Mountain Museum at Sulden to have a look at, and take a hair sample from, the second yeti in his collection. This was another trophy from the Schafer expedition, although displayed very differently. While the yeti at Juval that I had sampled the previous year was a stuffed head mounted on a wall, the Sulden yeti was a complete specimen enclosed in a display cabinet yet only partly visible. To add to the mystery, it was on a rotating stage and bathed in an eerie blue light.

Having received a call from Messner, the curator of the museum went around the back of the cabinet and unlocked it. With the rotating plinth turned off, I was able to climb inside, right up to the 'Snow Bear' as it was labelled. As there was no other illumination, the creature was still bathed in blue light. And what a creature it was. Standing about six feet tall, with one arm raised across its chest, it was like nothing I had ever seen before. Two amber glass eyes stared out from an extraordinary face that looked like a cross between a bear and a baboon, if you can imagine such a thing. It had a long sloping nose covered, like the rest of the face, with short pale hair. The region around the open mouth was heavily reconstructed with plaster or clay while the mouth itself held an assortment of teeth on both the upper and lower jaws. There were long Dracula-like incisors embedded in both jaws, but at the front of the mouth, where I expected incisors there were instead what looked like molar teeth – perhaps even inserted upside down. The rest of the body was covered in hair about four inches long, and seemed to me to be made up of parts of different animals. The hair on the arms and legs, for example, was considerably darker than on the trunk.

I tried to suppress a laugh. It was such a terrible fake that it surely couldn't have fooled anybody if they had a chance of close inspection. But a combination of the moving platform, the restricted viewpoint and, of course, the glow of the blue light made sure visitors had only a fleeting glimpse at each rotation.

Nevertheless, I carefully removed hair samples from six different parts of the creature for analysis. I wasn't expecting to discover a new primate, but this was a famous historical specimen so at the very least it was a rare chance to expose a Nazi practical joke.

18

The Explorer

One of the first responses to our call for yeti samples came from Christophe Hagenmuller, a French climber and explorer from the town of Annecy, not far across Lake Geneva from Michel's museum in Lausanne. That is where I met him and where he told me how he had come across his astonishing find. Christophe is a wiry, fit man who now works for an international software company that allows him time to indulge his enthusiasm for the mountains. He has been travelling in the Himalayas since 1996 and, like so many before and since, fell in love with that part of the world and with the people who live there. The way his eyes sparkled when he began his story showed this was no invented passion. His favourite part of the Himalayas is Ladakh, actually in India or, to be more precise still, in the Indian-administered part of Kashmir. Ladakh borders Tibet and many Tibetans crossed the border to live there after the Chinese annexed their homeland in the 1950s. Hagenmuller and a companion travelled to Ladakh every year between June and September from

1996 to 2003, walking with some Tibetan friends, one of who was from a Buddhist monastery. There was no particular objective in his travels. He just enjoyed being there among the mountains, getting to know the people, immersing himself in their culture and enjoying the nature of the place. Yetis were the last things on his mind.

'I was not looking for the yeti, I was just walking the area to discover the culture and looking at the nature, the flowers, the animals. I wanted to see the snow leopards – that was my goal but I was not focused on that. I just wanted to discover the nature and people. I learnt the Tibetan language there. I spent two weeks in a monastery where I was teaching English to young monks and in return I was taught the Tibetan language. During one of my travels, the second or third one, I was told about hunters who had killed strange creatures.

'It was an accident that led me to the yeti. One day we walked to a village. My friend was ahead of me and I saw a person sitting beside a horse. I asked him in Tibetan if he was okay and he said he was almost okay but he had fallen from the horse and couldn't go back home. I helped him getting again on his horse and went back to his house where he invited me to stay with his family for a day or two. And we started talking. I told him I was looking for the snow leopard and other creatures but he told me that he had something more interesting to show me. That's when he started telling me about the *tenmo*, which is the local name for yeti there. That's how I got introduced to the yeti, let's say.

'He hadn't seen the yeti himself but he knew that in his village a hunter had killed a strange creature forty years ago and thought he could show me the creature. Unfortunately after a few days and after meeting with the chief of the village we discussed about how I could see this creature, but finally they decided not to show it to me. They were afraid about what would happen if I would reveal the place and say what I had seen there. So I didn't insist too much and I said, "Okay, if I can see when I come back next year maybe, or in the future, that would be nice."'

Hagenmuller returned the following year, but once again the

village chief refused to allow him to see the creature. A year later, his fifth visit, a friend of his told him that he had heard that there was another village four days' walk away where a hunter had killed a yeti thirty years before. Hagenmuller continued:

'I had heard of the yeti, of course – a strange creature you can see sometimes but no one knows what they are. I wasn't focused especially on the yeti, for me it was more probably a bear or something like that which people would see in the dark or in special conditions where they were confused. I was interested but not that interested. For me it could be anything. It could be an animal like a bear or any strange creature, I had no idea on that before. But I thought I may as well have a look at this creature, so we set off for the village where the creature's body was said to be kept.

'We were travelling in a party of only three. The Tibetan monk who knew the way to the village, the man who I had helped coming back on his horse, and me. Only three persons. They didn't want me to bring any other persons to them and I had to promise that I would never reveal the place where I would see the animal. We travelled on a donkey for four days and, to be honest, I am not even sure I could find the village again. That is how we came to the village and then we went down the valley to a house. I have some pictures in mind of the house, and that's the one where we saw the hunter who had killed the animals, and we entered his house.

'It was very dark inside. In this area, because of the coldness, people have houses with very small windows. Some light entered the room but it was still really dark and I couldn't really see the animal. I asked if he could bring the animal outside, and he agreed. It was a sunny day so I asked if he could put the animal on the balcony or roof of his house, and he put it on the roof with plenty of light.

'I took a whole roll of photographs of this animal. Then, and I don't know why, I asked if I could take some pieces of its fur. I didn't know anything about DNA analysis, and anyway by then the creature had been dead for thirty years. The guy hesitated but said, "Okay, if you don't reveal where the animal is, you can take part of the fur." I put that in a little box where you put your film and that's how I brought the fur back to Europe.'

I had already seen one of Christophe's photographs of the animal. It certainly looked odd so I asked him about his first impressions of the creature.

'The first thing that came to my mind, which is funny, was that it was a mix of a wolf and a bear, two animals I didn't think could hybridise. That was the first impression. Then I looked more closely at different parts of the animal. I looked at the foot, the mouth, the teeth, and it seemed to be more a bear than anything else. But it was a very strange creature. I had seen a lot of bears in India as well as in the US and Canada and I wouldn't have said immediately that it is a bear, but I thought after some examination that probably it is a bear.'

We both looked at Christophe's photograph. The animal was about four feet in length with golden brown fur that was long and matted. It had a wide, flat snout, not like a bear at all, and a mouth with large teeth. Its front paws certainly had the claws of a bear, but it was the creature's head that looked distinctly un-bearlike. The ears, if they were there at all, were lying flat against the head, not sticking out as a bear's would, at least when it was alive. I asked Christophe if the man who shot the creature also thought it was a bear.

'No. He said it's definitely not a bear. He said, "I am sure it's not a bear and I cannot be confused because I'm a hunter. I've killed maybe thirty in my life and I can assure you it's not a bear. Don't tell me it's a bear. It is a *tenmo*." He wasn't hesitating at all. For him it was a *tenmo*. Period. When I started to joke about that he became a little bit angry like I was doing something wrong, not respecting his culture, whatever. Not respecting what he was saying seriously. He was not joking; he was serious, saying it was a *tenmo*.'

Having brought the precious sample back to France, Hagenmuller contacted the eminent palaeontologist Yves Coppens, an expert in many aspects of human evolution and a scientist of international reputation. They corresponded a few times but then nothing more happened until a colleague in Geneva saw the press coverage surrounding the launch of the Oxford-Lausanne Collateral Hominid Project in 2012 and mentioned this to Hagenmuller. And that is how the hair of the *tenmo* came to be in my laboratory, nearly forty years after the animal it belonged to had been shot. This was going to be a very tough sample to analyse. Eventually we got it to work and with a stunning result that we will cover later.

Hagenmuller plans to return to Ladakh quite soon. He thinks the hunter who shot the *tenmo* has since died. He wants to try once again to take a look at the first creature which he was denied on his previous visits. I didn't ask him precisely where the *tenmo* hair was found. It may have been technically correct to report an exact location, but far more important is Hagenmuller's promise to his Tibetan friends to keep it a secret. He may have seen and photographed a *tenmo*, perhaps the rarest of creatures, but he has yet to see a snow leopard, surely the most beautiful of all.

19

The Pangboche Finger

Cryptozoology is not short of good stories, but none beats the case of the Pangboche Finger. Fortunately there is a DNA angle, which allows me to include it without straining the boundaries of my enquiry. This case has everything. Skullduggery sanctioned by the most famous zoologist in Britain, tales of adventure by an intrepid explorer, the last of the 'Great White Hunters', some sharp moves by a Hollywood movie star, and all set against the snow-covered peaks of the Himalayas. And now some DNA results as well.

The story begins in 1959 when the director of London Zoo, the world-famous primatologist Professor William Osman Hill, summoned the Irish adventurer, Peter Byrne, already a veteran of the Texas millionaire Tom Slick's yeti-hunts in the Himalayas, to meet him in London. Osman Hill, alongside his regular research into comparative anatomy, had always nurtured an interest in cryptozoology, and yetis in particular. He offered Byrne a commission: to steal part of an ancient yeti relic, a mummified hand, from the Buddhist monastery at Pangboche in Nepal and to replace it with a human finger so the theft would not be discovered. Osman Hill pulled out the substitute from his office drawer and placed it on his desk. Byrne took the commission, and the finger, and set out for Nepal. With his long experience of the region, Byrne had little difficulty in crossing the border and trekking to Pangboche. Once there, he bluffed his way into the sanctuary where the hand was kept, snapped off the ring finger and replaced it, as best he could, with the substitute he had brought from London.

Peter Byrne told me the next part of the story when I visited him at his home in Oregon. While getting into Nepal was straight-forward, even with a dislocated human finger in his rucksack, getting the stolen digit out was much more of a problem. The theft from the monastery had been discovered and news had spread. Peter told me that he did not anticipate any great diffi-culty in getting the finger across the Nepalese border into India, but the customs in Calcutta were far more vigilant. Smuggling the finger back to Britain was going to be much more testing.

Then Peter had a lucky break. In Calcutta, he was staying at the five-star Taj Bengal and among the other guests was none other than the Hollywood actor James Stewart. Something I did not know about Jimmy Stewart, but Peter did, was that he was a very keen amateur archaeologist and anthropologist and a regular attendee at monthly meetings of the Malibu Anthropology Society in Los Angeles. This made the introduction easy. Peter Byrne explained his dilemma to Stewart and the most fantastic plot was hatched between them. Stewart and his wife Gloria offered to help, at great risk to themselves, and when they left next day for London the stolen yeti finger was hidden deep inside Gloria's lingerie case.

The Stewarts passed through customs in Calcutta without being searched but things did not go so well when they arrived in London, as Peter went on to tell me. The Stewarts were staying in their usual suite at the Dorchester Hotel on Park Lane. When their luggage was delivered to their rooms from the airport, Gloria's lingerie case was missing. Had it been intercepted? Had the finger been found? Anxiety turned to dread when, later that evening, there was a knock at the door and in walked a young customs officer carrying the case. Expecting the worst, the Stewarts offered the officer a cup of tea from the tray freshly arrived in their suite. The officer politely declined and instead asked Mrs Stewart if this was indeed her case, which she confirmed. Then he handed it over. Gloria saw at once that it was locked. If it had been searched, surely the lock would have been forced.

'It's still locked,' Gloria exclaimed.

'Of course, madam. The British Customs would never open a lady's intimate luggage.'

The Stewarts tried to hide their relief, and as soon as the officer had left clutching the autographed photograph that James had swiftly produced by way of thanks, the tea tray was put to one side and the drinks cupboard opened instead.

I caught up with the Pangboche Finger when my attention was drawn to an article in the *Daily Mail* in late 2012. Mathew Hill, a health correspondent for the BBC, had found out that the Royal College of Surgeons had discovered the finger while clearing out Osman Hill's archives, curated by the museum since his death in 1975. Hill arranged to have DNA extracted from the relic and sequenced by Dr Rob Ogden from Edinburgh Zoo who, according to the article, had found only human DNA. But what sort of human DNA could it be?

I went to see Dr Ogden in Edinburgh and he told me that he had indeed drilled out a piece of the Pangboche Finger and sequenced the recovered mitochondrial DNA. It was definitely human. I asked if I could have a look at the sequence, which he kindly gave me on a memory stick. When I returned to Oxford, I compared the Pangboche sequence to the many tens of thousands on my research databases from all over the world. Dr Ogden, sensibly given the age of the specimen (and who knows how old it was already when Peter Byrne snapped it off in 1959), had applied an ancient DNA approach that analysed the recovered DNA in short segments. Not all of them had worked and there was a gap of seventy-six bases in the middle of the sequence. In the runs of DNA sequence that Dr Ogden had managed to retrieve from the finger, I recognised some key variants. This was a European mitochondrial DNA sequence, in the clan of Ursula. The gap in the sequence missed out some important positions, but when I searched my database I found I could fill them in to create a rather unusual European sequence that I had only seen twice before. The Pangboche Finger sequence was almost certainly not from Nepal or anywhere else close by as it lacked the key variant (at position 16223) that is almost universal throughout Asia.

The DNA sequence recovered by Dr Ogden was of good quality across the regions that had worked and made sense by matching known mitochondrial sequences. But still the most likely explanation was contamination by human DNA that, as we have seen already, is the curse of ancient DNA work. I then began to wonder whose DNA this might be. It was unlikely to have been anyone from the Pangboche monastery. Theirs would have been a typically Nepalese or Tibetan sequence. It could have been one of the curators from the Royal College of Surgeons. It could have been Dr Ogden, but I soon ruled that out with a swab. Could it have been Peter Byrne's DNA or even Jimmy or Gloria Stewart's, the only three people that I knew for sure had come into contact with the Finger?

Often when I have analysed ancient bones I get multiple sequences from a number of the individuals who have handled the specimen. But the sequence from the Finger had come from only one person. I could tell that from the pattern of peaks on the output trace from the sequencing machine. They were all crisp and unmixed. Since I couldn't imagine the Stewarts handling the gruesome artefact more than absolutely necessary, the finger pointed at none other than Peter Byrne himself. It was he who had snapped it off in the monastery and smuggled it over the Nepalese border to Calcutta. So I made sure when I visited him in Oregon, that I got a cheek swab. And when the results came back, it matched the sequence from the Finger in every respect. Just a bit of fun, but also a demonstration of the persistence of DNA, which had remained on the surface of the Pangboche Finger for over fifty years.

Peter is a great survivor and so, clearly, is his DNA.

20

The Man who Shot a Bigfoot

In the Bigfoot world no recent case has aroused more controversy than the 'Sierra Kill'. Not only is the account gripping in itself, it has also stoked the embers of a long-running argument surrounding the deliberate killing of a Bigfoot. As soon as I heard about the 'Sierra Kill' I knew this was a most important case to put through the rigours of genetic analysis.

The story begins when Justin Smeja, an unassuming and like-able young hunter from Sacramento, an electrician by trade, was driving to his favourite hunting ground high up in the Sierra Nevada in northern California. I met Justin in San Francisco where he recounted his story, clearly not for the first time. He told me he had grown up hunting in the forests almost every weekend since he was a kid. Mostly he hunted deer, wild pigs, turkeys and bears. But not 'lions', by which he meant mountain lions, or cougars. It was illegal to kill a 'lion'.

In October 2010, he and his hunting buddy were in the Sierras of northern California at about 7,200 feet. They were on the lookout for bear and deer. They had seen a few deer, though none they liked the look of. With the exception of bear, Justin kills to eat, and these deer were too young. They headed for a familiar grassy clearing surrounded by pine forest, a perfect place for a clear shot at any deer coming out into the open to graze. As soon as they reached the edge of the clearing, Justin and his buddy saw the Bigfoot.

'So we drive in and look into this open meadow, and we see this creature that's on two legs. And at first glance I thought it was a bear, and me and my buddy saw it at the same time. I slammed on the brakes. And we're sitting there looking at this, and maybe for the first second or two I did think it was a bear, it was furry and was the right colour. I saw it was on two legs and that it looked kinda like a person in a bear suit or Wookie suit or something like that. (*Wookie was the huge, hairy, Bigfoot-like character from* Star Wars. *During filming in Washington State he needed an armed guard to protect him against Bigfoot hunters.*) It must have been seven or eight feet tall and at least 600 pounds. It was huge. I'm sitting there looking at it, watching it. A couple of seconds go by and I decide I'm gonna kill it.

'My buddy is yelling at me, saying stop, don't shoot, don't shoot, it's a person in a bear suit, stuff like that. I have a lot more hunting experience than he does so I decided it was my call, so I started squeezing the trigger. It started to turn to run away. It turned sideways and I shot it right here, on the top of its rib. It was a direct hit. It fell to the ground then started trying to get up and staggering around just like anything does when you shoot them. It started to try and get its balance and run away and that's when I could have had the kill shot. It was mostly staggering away on all fours, but just like if you shoot a person they don't usually run away on two legs, they crawl away fast.'

I hoped that this last remark was not taken from Justin's personal experience.

He was lining up for the second, fatal shot when his buddy yelled out that there were more of them. Justin took his eye away from the rifle scope and saw two smaller ones, maybe three feet tall and thirty or forty pounds. They looked like juveniles. Walking on two legs most of the time, sometimes on all fours, they melted into the wood. Justin and his friend got out of their vehicle and ran after them. They found them almost at once and, like the two hunters, the youngsters were looking for the fallen adult.

'We kept seeing them and they were obviously looking for their parent. It was just like when you shoot a sow pig: when you shoot a pig then you find out it has babies they always take you to the parent, so we were kind of following them, they were kind of following us and we were maybe fifteen feet away from them at times.

'Eventually I decided to shoot one. I was gonna shoot one from the beginning as soon as we saw one. I pulled up the rifle and my buddy is saying the same thing, no this isn't a good idea, we already have one on the ground, let's find that one and get out of here. So that was the plan, to find the big one and leave. Eventually I gave up trying to find it and said, "Let's just shoot one of the little ones, throw it in the truck and that way we'll have some proof that we can show people." So I shot the little one square in the neck, walked up to it, grabbed it. It was still alive, it was bleeding all over me.'

Justin immediately regretted what he had done, and for two reasons. When he held the dying juvenile close to his face, its almost human appearance panicked him. He thought he might have shot a human child, albeit a very strange one. The other reason was that they knew they had to pass a Park Ranger station on the way out. Realising the ranger would have heard the fusillade of rifle shots and was probably on his way to investigate, they decide to bury the body.

'We got a bunch of rocks and sticks and buried it, got in the truck, drove two and a half hours home without saying a word. Plan was to return the next day and get the body. That night there was a snow storm. It snowed 4ft and we were not able to get back up there.'

In fact the weather prevented them from returning for over a month. They eventually returned to the site in mid-November. The brought with them two trained cadaver dogs and Justin's bloodhound to help them locate the body. The dogs soon found the spot, but there was very little there. After eight hours of digging all they found was a piece of skin with hairs attached, which soon came to be known, rather fancifully given its diminutive proportions, as 'The Steak'. Though it was no T-Bone, there was plenty for a DNA analysis and a small portion was soon off to Dr Melba Ketchum's Sasquatch Genome Project in Texas. Four days later the results came back identifying it as a Bigfoot. Only later did Justin become suspicious that pretty well every sample tested by Dr Ketchum's lab had been identified as a Bigfoot, which is when he got in touch with me.

Justin has been back to the meadow and the surrounding woods over a hundred times since, but has never seen another Bigfoot. When I asked him whether he now regretted shooting the youngster, he replied quite calmly that his only regret was not to have put the body in the trunk of his car and driven it home. Then he would really have had something to show people. I got the impression talking to him that on the fateful day he did not realise quite how precious a Bigfoot body would have been. He had certainly heard lots of Bigfoot stories, but until the 'Sierra Kill' incident he was not particularly interested in them. Though he had spent most of his life hunting in the woods, until the day he shot one he had never seen any signs of Bigfoot and didn't really believe they existed. Like many eyewitnesses I talked to, Justin is now driven, almost to the point of obsession, to convince others that what he saw was real. I had the impression that his desire to be believed was even more important to him than the immensity of the discovery had he been able to produce the genuine body.

As Justin finished telling me his story, he produced a small envelope. In it was a sliver of skin about an inch long by a quarter wide, bone dry now, but still with plenty of short almost blond hairs attached to it. This was all that remained of 'The Steak' and he handed it over to my care. Justin told me he had stored 'The

Steak' in salt to preserve it, hoping this would not have harmed the DNA. On the contrary, I was able to reassure him, it would have helped. Salt preservation is one of the reasons why DNA from Egyptian mummies survives so well. I immediately put 'The Steak' and the envelope into an evidence bag. He also produced his hunting boot, the one that had been spattered in blood when he held the dying juvenile. At first I was reluctant to accept the boot for DNA testing as unlike hair, which I knew I could clean up before the analysis, the blood on the boot was bound to be highly contaminated. But I also realised that the connection between the blood on the boot and the dying Bigfoot was much more solid than it was with 'The Steak', which had been recovered weeks later and may have had nothing to do with the creature that Justin killed. Luckily I carried a scalpel in my sampling bag, and with Justin's help to identify the exact spot where the blood had landed on the boot, I cut a sliver from the surface and placed it in another evidence bag.

I must say I was slightly surprised that Justin had the time to notice precisely where the blood had landed while the juvenile Bigfoot was choking to death in his hands. Not long after taking 'The Steak' and the blood sample from Justin's boot we knew a great deal more about their identity, as we shall see.

21

The Veteran

Dan Shirley, who like Justin was also from Sacramento, had picked up news about my Bigfoot research project through the Internet. We arranged to meet in the lobby of the Airport Marriott Hotel in San Francisco, a sampling venue I had used before for my book *DNA USA*. He arrived with his research partner, Garland Fields, and both of them looked intense. Dan wore a red sweater with a swooping bald eagle wrapped in the Stars and Stripes, yellow talons outstretched. His long greying hair was swept back and contained in a patterned blue bandana. He could have been a biker, a Hell's Angel, and I was half expecting to see a shining Harley propped outside. But mostly he reminded me of 'The Boss' and I thought I heard the opening bars of 'Glory Days' by Bruce Springsteen playing faintly in the background. Dan and Garland looked me straight in the eye. When I asked Dan about himself, what I heard did nothing to dispel my unease.

Now working in what he called 'private security', Dan had fought in Vietnam back in 1972 after training in special operations. He didn't say much about his time in the jungle, only that he was in an ambush squad.

He really did say, 'Something happened in 'Nam,' but that was about as much information as he volunteered. I was keenly aware that I was pushing my luck when I asked him whether he had killed anyone.

'Yeah. Oh yeah. It's a do or die situation. Oh yeah. You betcha.' That was as far as Dan wanted to go. He continued:

'From that point it took quite a few years of adjustment, and I just basically went along trying to live a normal life. I always liked being in the woods or in the wilderness. That never left me. When I got out that's the first thing I did. I went right back to it, and I'm still at it.'

Dan first heard about Bigfoot in 1967 when the Patterson-Gimlin film of the Bluff Creek sasquatch was front-page news. Soon there were Bigfoot sightings all over California, even close to his own hometown, Roseville, a few miles from Sacramento in the far-from-wild Central Valley. He was around fourteen when he went up to Bluff Creek, where the Patterson-Gimlin film was shot, with his father, a lifelong believer in Bigfoot. That's when the bug bit him, and it has never left. Dan and Garland have kept away from the agitated world of Bigfootology. You won't find them at any of the dozens of meetings where the enthusiasts get together to exchange the latest information, or just to gossip. They work alone. Together, but alone. And they work systematically. Over the years, Dan and Garland have identified several 'hotspots' in California and southern Oregon where they have either seen a Bigfoot or had a close encounter with one, by which I mean heard or smelled them, or been pelted with rocks.

To attract a Bigfoot their method is to hang apples as bait (always green apples – they work best) from a tree branch about 7–8 feet from the ground. On the trunk of the tree Dan smears a special grease to catch any hairs the Bigfoot might leave as it reaches for the apples. He won't tell me exactly what's in the grease, only that it is based on a mixture of four or five different animal fats. He puts it in the freezer until it hardens. It has taken him years to perfect this method and in a particular hotspot in the Sierras over the past year he has baited the same tree thirteen times. Only once was the grease unable to capture a hair. The apples were always taken. After he has baited the tree and smeared grease on the trunk he always indulges in some wood knocking before he leaves. Like the formulation of the grease, his particular routine has been fine-tuned over time.

'It's taken me three years to perfect certain segments when I'm wood knocking. I try different sequences of wood knocking. Basically I'll either use two sticks and knock them together or I'll do it in the natural form and knock on a tree. You'll get more results with the natural knocking on a tree. But there seems to be a particular way that you do wood knocks and it's how many knocks and segments that counts. You'll do like "boom boom boom", and maybe a "boom boom boom boom". And I've come to find out in my research it's almost like if you're playing the guitar and looking for the right chords.'

Dan is a great believer in wood knocking for communicating with Bigfoot. If he gets it right, then he will often get a knock back in response. This can even develop into a 'conversation' lasting several minutes. He disdains the alternative method of calling, often claimed by other enthusiasts to be the more reliable way of getting a response.

'From my personal experience I have not had any response whatsoever, with what they call a Bigfoot yelp or a Bigfoot howl. I found it kinda fruitless, I've never had a response. The only response you normally get is if you're in an area where there happens to be wolves or coyotes. That's about all you're usually gonna draw. And of course I know people that's had these particular sounds on audios, but it doesn't excite me at all because of the fact so often it's obviously wolf or coyote, and some cases even elk. Up here where we're at, number one we have no wolves, no elk, no moose, but we do have coyotes. I've done the yelps and the howls and to be honest with you after a while I kinda felt pathetic.'

Over the years they have worked the 'hotspots' together, Dan and Garland have had two clear daytime sightings.

'The first time, we were up in northern California, almost into Oregon, and were working one of our research areas which we call Rock Creek Road. On one particular day about ten in the morning we were heading up to our research area. As we pulled around there's a road that comes to a bend. When we came round this bend I looked up the side of the mountain and right on top of the ridge among some pine trees I seen this big black object. I stopped the vehicle and I got out and I started walking down the road so I could get a better view. I thought it was a black bear. As I started getting closer to it, I never left my eyes off it and as I'm looking and observing saying well, that's got arms. It's got hands, it's got a head. And it's the way it was standing next to the tree. And it was black and you could see it was all covered in hair. And then it disappeared.'

The pair climbed to the top of the ridge, but they never saw the creature again. The second clear sighting was in another research area called Mosquito Ridge in the Sierras above Roseville.

'That was in the winter. Me and Garland went down by the American River, on what we call the North Fork. We went down to the bottom of the canyon and got to the river and we were sitting there observing. To my left this object caught my eye and when I turned around and looked this squatch was walking down the canyon. Just strolling down like it was nothing. You could see it going from tree to tree as it's coming down and it's got that swagger type walk. And at the same time I'm thinking you gotta be kidding. Because I got a whole total view of him, at first I thought it was a human all dressed in black. Garland was to the right of me and I says, "Garland, look at this," and he turned around. "Look," I said. "What?" I said it looked like a guy walking down the Canyon, big tall guy dressed in black. Garland said "I didn't see nothing." I said, "Wow. Strange."

'And then I looked again and the squatch popped up again and went behind the foliage. And I told Garland, I said, "Hey, that ain't no man; I'm telling you right now that ain't no man."

Where would they come from? They're in the middle of nowhere, that ain't a human. At that moment we're both looking, and all of a sudden this thing stands up. The foliage is kinda high but he stands up and all you see is this great big huge pair of arms coming out and you could see right here, and you could see the hands. You could count the thumb and the fingers. He was probably about sixty yards away. I grabbed the camera and when I opened up my cam to record it, my camera wouldn't work.'

I asked Dan what sort of creature he thought the Bigfoot was.

'I firmly believe in the *Gigantopethicus* theory. I lean more to that. It's also a high possibility that maybe a particular classification of Neanderthal still exists. One thing a lot of people don't realise is the fact that the particular woods where we're at is prehistoric wilderness. It hasn't changed for thousands of years. There's an area up here that man has never even walked in and that's the truth. Haven't even seen. I would love to come across that.

Earlier, in his email to me, Dan mentioned that he and Garland had recently found two hair samples from another 'hotspot' research area near Grass Valley in the Sierras, so I asked how he came across them.

'That particular day me and Garland were gonna do our research in that area. I wanted to park on top of the hill and start walking in, because I wanted to try to catch them off guard. We were walking down that particular mountain road that leads us to the entrance of the area we go into. There's a fence line that goes alongside to the left of the mountain road and I always look at the fence line for any kinda maybe evidence might be hanging in the fence, that type of thing. There was nothing. I also look for footprints, as we find a lot of deer and bears. We've also got a couple of bobcats up there, a male and a female running around same as usual.

'We're at the research area about three or four hours and as we were baiting the tree we found some hair that was on the tree, on one of the limbs. I was thankful for the grease I put on there. It had got kinda gummy and sticky and as they're going up reaching for the apple so we grab some hair from under the arm. And so we found the hair on the bait tree, so we diligently pulled it off the tree using tweezers and put it in our sample bag without touching it; there was absolutely no contamination. We set our trail-cam and headed out. We're going up to the road and the fence line is to the right of us. When I get to that particular area I stop because something caught my eye, and I'm like what's that? So turn around I look and there's this beautiful set of hair just hanging on a barb.

'We went up to the hair to investigate it, we checked it out. As soon as we looked closely at it we both knew it hadn't come from anything in these woods. No bear, no deer, no coyote. At the same moment right across the road, down a little gully with some thickets and a lot of foliage, we sensed there was a sasquatch. My guess is it tried to hurdle the fence and its inner thigh got hung up on the barb. We could smell the squatch. Like it was right in our face. The squatches up there they have a really strong intense skunky pungent smell. It was probably between eighty or a hundred feet from us and smelling really bad.'

Dan explained that it was his 'sixth sense' that had warned him that the Bigfoot was close by. This sense, which Dan thinks we all have but rarely use, was heightened during his time in Vietnam and had saved his life more than once.

'One thing that I know about human beings is, we're no different from any other animal. Animals have a sixth sense, we have a sixth sense, but the problem with human beings is that most humans are not forced to use their sixth sense. When you're in situations and you're traumatised to a very high point, you're actually forced to dwell in your sixth sense that we're all born with. And of course in combat or in war itself you're living those situations. It's continuous. You become an actual animal yourself and by doing so you start tapping into your own sixth sense. And once you have tapped into this sixth sense, you've tapped into a complete area that dumbfounds human beings. They don't understand that aspect because they've never had to be pushed to tap into it.

'I get a sensation like we were talking about when I know the creature's there or not there. Or sometimes I get premonitions where I can see what's ahead of me, though I'm not there yet. Once I get up in the woods, it's an automatic click with me, like hitting a light switch and my sixth sense is popping automatic. Now when I'm there I'm with the animals and with everything in the woods, I become part of it. Of course it helps with tracking skills. I get a better sensation of smell, hearing, eyesight. And it's been a big aptitude to my research, big big aptitude. In fact it gives me an edge.'

That day, Dan's sixth sense told him there was a Bigfoot very close by. He and Garland were walking down a path, not far from Mosquito Ridge, and they had stopped to rest when he got that familiar feeling that he was being watched.

'I decided to look into the bushes. I walked up and looked through the foliage and I tell you what, I jumped back about ten feet, and so did Garland, we both were like "Woah!" There was a big male just on the other side. The only thing that kept us separated was the foliage. I mean, the only thing. And I think what happened was I'm peering in at it and I don't see nothing but they see you. Then all of a sudden he jumped up and just threw his body weight and you know this thing was huge, it was like a thousand pound grizzly bear ready to break the brush, coming right at you. So I jump back and actually I went for my gun, because I didn't know what it was gonna do, I had no idea.

'That's the first time we experienced when they have gotten aggressive, but I will say this, if you do experience such a thing, it'll scare the hell out of you. I don't care how long you've been doing this, if you have an interaction like that it'll scare the hell out of you. But in reality I think it was as startled as we were. When our eyes met, they've usually got the upper hand and in this instance he's got eyeballs looking at me. It was a hell of an experience, I'll tell you that.

'We go in the wilderness a lot and the only reason I carry a gun is for self-protection. Garland carried a rifle in those days and I carried my sidearm. I kinda felt like I got to the point where I think we could get a lot better interaction and maybe they wouldn't be so aggressive if we didn't carry the rifle no more. I usually wear clothing that hides the '45 so they can't see it. And once we started leaving the rifle behind, they seemed to be not aggressive towards us anymore. It's no good to be running around with guns and trying to hunt these creatures.'

The squatch by the fence didn't show but Dan didn't think it was very happy. They decided to retreat, but not until they had worked the hair loose from the barbed wire fence, bagged it and tagged it.

Dan reached into his rucksack and pulled out a Ziploc bag. Inside was a tuft of light blond hairs. I put the whole thing into an evidence bag and within the hour it was in a FedEx box and on its way to the lab.

I need not have been nervous meeting Dan and Garland. They were both charming and enthusiastic. Whatever it was that happened to Dan in 'Nam, it is well behind him now.

22

The Landscape Gardener

On a clear day, look south from Seattle and you'll see the snow-covered volcanic cone of Mount Rainier reaching 14,000 feet into the sky. Shift your gaze ninety degrees to the west and at the same distance of fifty miles a range of snowy mountains appears on the horizon. These are the mountains of the Olympic Peninsula. Unlike the comparatively isolated Rainier, the Olympic Range is a wilderness of high peaks and ridges separated by thickly forested, steep-sided valleys. There are roads around the edges, but none that penetrate the interior. Ideal sasquatch habitat, one might imagine.

I was eager to meet Derek Randles, who knows this area as well as anybody – not just because of the high regard in which he is held in the Bigfoot world, but also because he is someone whose whole life was changed in an instant by what he saw in the woods.

'It all started for me in August of 1985. Myself and two friends of mine were hiking up to the Olympic National Park up by Lake Kushman. We had gotten a late start that day and were about six miles up trail. We were all in our twenties, all in top shape and hiking fast. It was getting a little late in the day and we got to the finger ridge we wanted to explore. We were losing light and we wanted to get up at least to a level place to make camp so that the following morning we could get up and do some exploring, maybe a little rock climbing, something of that nature.

'We found a level meadow area with a timber line about forty yards over from us. I had this habit of carrying pruning shears in my backpack and I would go around to some of the trees and lower branches and prune off some of the limbs and make myself a bit of a mattress under my tent. So I'm in the process of doing that while Mike and Jim are taking off their backpacks and we're just settling down.

'We just got started when we hear this crash. And it stopped us in our tracks. We're literally in the middle of nowhere, and heard this big thunk. We all looked up to our left and I thought that, okay, maybe it's an elk taking off or something. Mountain goat, something. And as we're looking up in this area suddenly this rock comes arching out of the timberline and it's probably between a softball and volleyball sized. It's not falling off a cliff because we're on top of this finger ridge, and there's no cliff above us. The rock was thrown and it actually arched, came down and landed to our left. You can't wrap your mind that fast around something like that, especially given where we were. This rock hits the ground and we all look at each other like "What in the world?" and it was very scary because it was a good-sized rock.

'All of a sudden a second rock comes and that's when we started getting very scared. The second rock landed very close to the first one, 15 feet off to the left of us. A resounding thud on the ground and we said, "We're out of here." We start picking up our gear, and another rock comes in. Now we're extremely panicked and it's just about dark and that's adding to the fear. As we're about to get going, another rock lands about ten feet to our right.

'Now we're just scared, and we're gathering our gear up as fast as we can. As a matter of fact we didn't even take time to put our backpacks on. We just grabbed them off the ground and started running down this ridge. And as we're going I got about ten steps into it and remembered I had a 357 in my backpack (*a .357 Smith and Wesson Magnum – a very powerful sidearm*). I stopped quickly and my two buddies kept going down, didn't even realise I'd stopped, and I reached in my backpack real quick to get the gun.

'I looked back up the hill and there it was. It had come out of the timber line and was just standing there in the middle of this meadow looking down at us. It was terrifying. Here's this thing that's probably eight feet tall covered in hair and standing there swaying. And my mind just went berserk. There's a sasquatch standing there looking at me. It just stepped from myth right into reality. It seriously changed my life forever. It altered the course of my life and honestly within four or five days of that event I went into research. That was twenty-eight years ago and I've been in Bigfoot research ever since.'

Of course, I wanted to know what the Bigfoot looked like.

'I got a very clear silhouette, a very clear one because it was standing out in the open at this point. The light had faded to a point where I couldn't make out facial features, maybe a little bit, not really but the silhouette was very clear. And it looked like your classic sasquatch standing there. I would guess somewhere around eight feet tall, I'd guess somewhere from eight hundred to a thousand pounds. It's very big, its arms were very long, it had muscles. It looked like a bodybuilder. Not like Arnold Schwarzenegger but you could just tell it was well defined and cut and ripped and big. I saw the silhouette really well and I could see the outline of the hair. There was no mistaking what it was. It wasn't a bear, and it wasn't a person. It was very large, standing up on two legs, standing there swaying looking at me.'

Jeff immediately discounted the suggestion that it might have been a big grizzly standing on its hind legs, or maybe a person in a costume.

'No, it wasn't a grizzly; there are no grizzlies down here in the Olympic National Park. To say there's a sasquatch and no grizzlies sounds kinda crazy but no it wasn't a bear. I've been around wildlife – bears, deer, elk and so on – my entire life and I've encountered probably close to a hundred bears along the trail. I am a wilderness guide here in Washington State and I encounter bears very frequently. I know exactly what they do and this was not a bear. It was more ape-like, very big and the shape of a human, but very muscled up, very large, incredibly thick and just incredibly big and heavy.

'Could it have been a person? I highly doubt it. If this was a person they would have been massive. Also this person would have to have used extrasensory perception to know where we were gonna be. We didn't even know where we were gonna be that day. There's no way somebody would already be pre-positioned up there in this incredible outfit, eight feet tall, very heavy, just waiting for us to come to the spot we didn't even know we were gonna go to ourselves. So there's just no way. Absolutely no way. I know what I was looking at.'

Eventually Derek and his friends, running through the dark, reached an established campsite. Even so, Derek had the distinct impression that the creature was following them, not on the trail but parallel to it, inside the forest. Derek was the only one of the three to have actually seen the Bigfoot. The others only heard the crash and saw the rocks being thrown. As very often happens, Derek did not own up to the sighting straightaway for fear that his friends would think he was crazy, even though they had been close by. The next day they hiked back down to a Ranger Station and Randles told the ranger about the events of the previous night. He just got a funny look and was told someone would contact him in a week or so. They never did.

Derek went back up to the finger ridge meadow a month later, by which time the snow had started to fall. He didn't see or hear the Bigfoot but he did find a set of tracks in the snow and measured the prints at sixteen and half inches, extra-large even by Bigfoot standards. The whole experience did indeed change Derek's life. He quit his job and set up in business as a landscape gardener so that when he got the business established he would have the time and would be in the right place to indulge his passion for these creatures.

Nearly thirty years later he has lost none of his enthusiasm for Bigfoot, despite never having seen one again. He has found eighteen sets of tracks and felt that he was in the presence of a Bigfoot at least a dozen times, but he has not seen one.

As well as running training camps, where clients learn about the practical aspects of Bigfoot research, such as casting footprints in plaster of Paris, Derek also runs the 'Olympic Project' which has the declared aim of providing both proof of existence and a species identification for the Bigfoot in his area. His main weapon is an array of trail cameras, fifty-six in all, and his intention is to saturate a small 'hotspot' area until he gets a good image on film. The trouble, he admits, is that nowadays photographic images are so easily manipulated and there have been so many hoaxes that he suspects that even if he succeeded in getting a good image of a Bigfoot, many people would still refuse to believe it was genuine.

As far as DNA goes, he doesn't think it will provide species verification, though he admits he could be wrong about that. Which he is, but only if the study is carried out properly. He had already given away a lot of hair and other material to Dr Ketchum's Sasquatch Genome Project, which I covered in an earlier chapter. However, not all his samples had gone to Texas and he had two left, which he contributed to the Oxford-Lausanne project for DNA analysis.

The first hair sample came from Harstine Island in south Puget Sound, off the southeast corner of the Olympic Peninsula. Back in 2009, around Christmas, a local family had reported a couple of Bigfoot sightings on the island. At the time Randles was working for another Bigfoot organisation which co-ordinated reports of sightings and sent staff to investigate and to interview eyewitnesses. To Randles, the family from Harstine Island seemed pretty reliable, so he stayed in the area hoping to see the Bigfoot for himself. Over the course of the next three days he got some really loud vocalisations and on the fourth day, which was the day after

Christmas, he discovered over a hundred prints in the snow only a quarter of a mile from the witnesses' house. He followed the tracks and they led through some blackberry bushes, which in the Northwest are especially thorny, according to Randles. Snagged on thorns seven feet above the ground he found strands of black wavy hair. Randles believed that there was a high probability that they were from the Bigfoot that left the prints. Being found so far above the ground they could not, he argued, belong to a bear. Randles passed these hairs on to Dr Jeff Meldrum, who sent them on to me.

The second sample came from the west side of the Olympic Range at another of Randles' study sites. It was not a habituation area, like Dan Shirley and Garland Fields' from the previous chapter, where baits and so forth are laid over a period in an effort to encourage the Bigfoot within range of a trail-cam, but it was an area where people have had classic sasquatch encounters with vocalisations, sightings, wood knocking, the lot. Here, a couple of investigators with the Olympic Project doing thermal imagery work started finding hair stuck to some of the trees. Why the hair was stuck to the trees they were not really sure. It could have been a scratching post but, in any case, they began to gather the hair. A lot of the hair went to the Ketchum study and she reported back that it passed all of her criteria for sasquatch.

Fortunately, Randles did have a few hairs left which, again through Dr Meldrum, were sent to Oxford.

23

The Indian

Marcel Cagey is a member of the Lummi tribe whose lands are on the Pacific coast of Washington State, not far south of the Canadian border. He is involved in some capacity in the luxurious Skagit Valley Casino Resort, which is owned and run by the Lummi. Rhett Mullis has been going to the Lummi peninsula for several years, following up the frequent sasquatch sightings that occur there. He has been patiently cultivating the tribal leaders and getting them interested in what must be one of the most productive sasquatch sites in America, judging by the number of reported encounters. That surprised me when we drove up from Seattle, over the Skagit River and onto the Lummi tribal land. For a start the landscape was completely flat, though the sunlit peak of Mount Baker was not far away on the eastern horizon. A pair of handsome bald eagles flew overhead, heading for the mudflats exposed by the retreating tide to join others perched on stranded tree trunks carried down by the river.

Marcel's home overlooked the bay. Like most of the others

spaced out along the road, it was a functional single-storey dwelling, certainly not fancy. Outside, however, were parked an enormous yellow Hummer and a new Mercedes. Marcel, a man in his thirties with thick black hair cut so as to leave a crest on top, came out to greet us. He made us feel very welcome, but frankly the house was a mess. What looked like rubbish strewn around the garden, Marcel explained, was actually a gift to the local sasquatch that came into his property to feed. We headed round to the back garden as Marcel pointed out a gap in the hedge through which the squatches came and went. It certainly had the trampled look of a game trail, but I couldn't find any hairs clinging to the shrubs on either side. In the trees beyond the hedge was where the sasquatches lived. Lots of them, we were assured.

The woods didn't look very promising. They were open, mainly deciduous and, according to the map, didn't stretch all that far. Marcel then began a ritual to call in the local squatch. He told us that he had already done a little praying and singing to welcome them in and bless the air before we all got together. He had assembled a lot of expensive equipment to help him with a microphone and a bank of speakers ready to broadcast his calls far and wide. Like the Vietnam veteran Dan Shirley, he favoured wood knocking over howling as a way of attracting the squatches' attention. But there was another important element in the ritual. Marcel took out a sheaf of dried Californian desert sage, lit it and hung it on a line from the branch of an apple tree. The air was still, and soon a column of sweet-smelling smoke curled up into the evening sky, white against the setting sun. As the sage swayed back and forth, Marcel began his incantations in a language I could not recognise. A few minutes later he took a stick of wood and knocked it against the apple tree. The amplified sound boomed from the speakers and out into the darkening woods. Then Marcel explained what he was doing, and why the sasquatch was so important to him.

'I like giving back to it. It's never harmed me yet because I respect it and I don't go looking in the woods much anymore because of this respect we both have. I like singing it songs and welcoming it to my life. Thinking how it's touched my life spiritually. I'm just thankful for that. Thank you for allowing me to share all these great teachings and things that you have shown me, it's kept me on my right path.

'It did change my life 360 degrees. Two and a half years ago I was pretty much like any American, out for making lots of money, and you're not happy till you have all the money in the world I guess. But when I had this creature come to my house it really shook my cultural beliefs. It made me understand who I am. Kind of like a message from the Creator. "Money's not for you. This is for you." You know because natives struggle so much in this world, in this life. All we got is our spirituality to let us know we're going to be all right. So this creature has been very spiritual for me. It changed my life. I lost a lot of people before. You know I just lost my mother just before that creature came here.

'It's almost like a message. Don't go after that world, son. Stay who you are. I had to take a step back and look at my beliefs, because a lot of these teachings are here but I was never there to receive them. It kind of shook my world and I need to be in contact and thank him for restoring my culture. I was being assimilated, I was going through assimilation, I was shedding my native ways and my teachings. I was becoming like all the other people out there. Being successful and trying to provide for their future by owning all these things . . . just corporate America, you know?'

By the time Marcel had completed his ritual, it was getting quite dark. I had been scanning the unkempt lawn for some of the sticks Marcel told us the squatch was in the habit of throwing into the garden. I had seen a short one about eight inches long, but I didn't want to move while Marcel was in full flow. While he and Rhett were talking I walked over to the stick and picked it up. It was too dark to see if there were any hairs caught on its surface, so I took it over to Rhett's car, switched on the headlights and held the stick in the bright beam. Most of the bark had been stripped away, but trapped under the small fragment that remained was a single very fine hair. It was dead straight, about three inches and a light red-brown colour. When I showed the stick to Marcel he immediately confirmed that it had been thrown onto the lawn by the sasquatch very recently. It was typical of the gifts he had been finding for several weeks now and he assured me it did not belong to the apple tree or anything else that was growing in the garden. Fetching my tweezers, I gingerly removed the hair from beneath the bark. Being careful not to drop it and lose it for ever in the grass, I slipped the sasquatch hair into an evidence bag ready to go to the lab.

The following day, Marcel and Sam, a Lummi neighbour, took us on a tour of the sasquatch hotspots on the peninsula. We saw plenty of evidence. A red cedar stump that had been gnawed by a squatch going after grubs living under the bark. Tree limbs bent and twisted in a characteristic way as the creatures stomped through the undergrowth. Impressions in the ground that were surely giant footprints. The place on the road where the school bus driver had seen a squatch cross only yesterday. A river bank from which Sam had clearly seen three of the creatures on the opposite bank a month before while he was fishing for salmon. Everywhere we looked there were signs of sasquatch, if only you had the eyes to see them. Not being blessed with this facility, I clutched the evidence bag a little tighter.

24

The Government Laboratory

My invaluable and indefatigable guide to the world of Bigfoot, Rhett Mullis, had arranged for us to meet up with two hunters who had experienced close encounters with Bigfoot. We met in the town of Medford, Oregon, just north of the border with California, and right in the middle of Bigfoot country. To both the east and the west of Medford richly wooded slopes lead up to the volcanic peaks of the Cascade range that had burst through the earth's crust eons ago.

Our rendezvous with the hunters was at Elmer's Restaurant, one of many diners in Medford where locals gather to eat themselves into an early grave, under the benevolent eye of the founder Walt Elmer whose portrait hung on the far wall. One of our companions, Greg, was larger than life in every way, while Tom, his hunting partner, wiry and shy, was the complete opposite. Despite their apparent personality differences, Tom and Greg had been hunting buddies for many years, tracking and shooting deer and elk, especially in the densely forested flanks of the mountains to the south and east of Medford. I set my voice recorder as Tom began.

The first time Tom saw a Bigfoot was on one of these hunting expeditions, this time in the woods just over the state boundary into northern California. It was broad daylight as he and Greg climbed to the top of a ridge and looked down into an unfamiliar canyon with a small river at the bottom. They decided to try and flush out any game that might be in the scrub surrounding the river and began to inch their way down the steep wall of the canyon. When they were nearly at the bottom, they heard a crashing noise, the sort you might hear when a large animal is breaking cover. Expecting to see an elk racing out into the open, they stopped and waited. A few seconds later, a creature on two legs appeared at the bottom of the opposite slope and in no time at all bounded up the side of the canyon and disappeared over the ridge. Amazed, Greg and Tom wondered what they had just seen. The creature was about six feet tall, with almost white fur, long arms and very powerful, muscular legs. It had climbed the steep canyon slope, which was about a hundred feet high at that point, in a matter of a few seconds. It didn't seem to be hurrying, just going very fast.

Tom and Greg looked at each other, trying to make sense of what they had just seen. I asked them if it could have been a bear. 'Not a chance,' said Greg. Like all the woodsmen that I interviewed they knew perfectly well, from long experience, what a bear looked like and how they behaved and moved. Crossing the river, Tom followed the same route as the creature up the opposite of the canyon. Fit though he was, Tom found scrambling up the loose rock tough going. It took him about three minutes to climb the hundred feet to the ridge. The creature took no longer than ten seconds.

Like everyone in that part of Oregon, Tom and Greg had heard the stories about Bigfoot and took them with a large pinch of salt. That was until the moment when they saw one for themselves. I asked about the white fur, which is not normal for a Bigfoot. Apparently, so Greg told me, there had been other sightings of an albino Bigfoot a hundred miles to the east towards Idaho. If it was the same creature, it certainly moved around a lot.

We moved on to another Bigfoot encounter in the same general area the year before. Greg and Tom were in a part of the forest where a mountain lion had recently killed a jogger. There are a few thousand 'lions' in Oregon, but they are very rarely seen. They hunt mostly at night by ambush, lying in wait near a game trail then pouncing on any animal that is passing. Their prey is mostly black-tailed deer, which are everywhere in that part of Oregon. However a lone female jogger had been ambushed in broad daylight on a trail. The consensus was that the mountain lion would have been stimulated to attack by the fact that the jogger was running away. She would not have heard it coming and only when it sank its fangs into her throat and throttled her would she have known what was happening. Greg and Tom were in the area to see if they could find the mountain lion.

Greg's encounter with the Bigfoot happened at night. He was in a thickly wooded part of the forest where tall Douglas firs stretched a hundred and fifty feet and more into the night sky. It was very dark so he was equipped with night-vision glasses, which pick up heat from live animals and relay a ghostly green image through the eyepiece. The undergrowth was a mix of shrubs, ferns and young trees. Greg heard a twig snap some distance behind him and quickly turned round. Through the night-vision goggles he could see the terrifying image of a large creature approaching at speed. There wasn't enough definition to make out any detail, or to tell what sort of creature it was. Greg quickly drew his rifle, and without thinking, instinctively removed the goggles when, of course, he was plunged into complete darkness. He could hear the creature coming towards him, but he could not see it. The noise of its approach through the undergrowth grew louder and louder and Greg raised his rifle, but could not see a target.

The noise stopped and Greg became aware that the creature,

whatever it was, was right up close in front of him. He could feel its warm, fetid breath on his face. He could still see nothing. Afraid to move lest he be attacked, Greg stood rooted to the spot. The creature made no sound and just carried on breathing into his face. After what seemed like a very long time, but may have only been a matter of seconds, he no longer felt its hot breath. There was a rustling in the undergrowth, and the creature was gone. The next day Greg and Tom returned to the spot and found the outline of a large footprint in the mud beside a shallow pool. A true entrepreneur, Greg has capitalised on this experience and now runs a Bigfoot adventure company, taking clients into the woods at night in the hope of a similar encounter.

Some way into our conversation, Greg threw in the casual remark that there were three Bigfoot hairs stored at a laboratory nearby. I immediately asked for more information. His former wife, Greg explained, is a hairdresser and one of her regular clients was the wife of the lab director. Greg didn't know much about the lab itself, except that it was located in the nearby town of Ashland, ten miles south of Medford. This was obviously something we should look into and within minutes Rhett had googled the lab, found the phone number and dialled it. Rhett spoke to the operator and handed me the phone. To my delight, I was immediately put through to the lab director, Ken Goddard. Though not prepared, I must have managed a good enough introduction to myself and the project to be invited over to the lab later that morning.

Within two hours, Rhett and I were driving through the security gates of the US Fish and Wildlife Service Forensic Laboratory in Ashland. By then I, or rather Rhett and his smartphone, had found out that this was no ordinary laboratory, but the only one in the world exclusively devoted to wildlife crime. Talk about serendipity. On top of the coincidence that Greg's ex was the director's wife's hairdresser, we had stumbled on the only government laboratory in the world that just might have a genuine interest in what I was doing. And it was only ten miles down the road. But the good fortune didn't stop there. In my experience, government lab directors are a guarded bunch, so the most I expected was a few minutes of polite conversation in the reception area. But Ken Goddard is no ordinary lab director. As soon as we passed through the security gate, he was there to meet us. We were invited in and were given a tour of the whole facility.

The lab's remit, Ken explained, was to provide the forensic backup for wildlife crime investigations, not just in the US but, through international contracts, in over 150 other countries around the world. As we toured the facility, which was in itself a rare privilege, we saw cases of elephant ivory, seized by customs officials on their way to China. Their origin was being investigated by the latest techniques of forensic analysis. Intercepted on the way out of China was a large cardboard transit box labelled 'Special Quality. Rhinoceros and Antelope Horn Febrifugal Tablets' which had been seized en route to supply the alternative medicines market. In the dissection room a black bear was undergoing an autopsy to see of it had been poisoned. In another room were the skulls of a dozen tigers, all with a neat bullet hole right above the eye sockets. Ken explained that the elderly tigers had been retired from zoos and handed over to the custody of what, on the face of it, was a genuine charity. But genuine it certainly was not. It was, in fact, a sickening racket uncovered by Ken's agents. For money, clients shot the defenceless tigers at point-blank range in their cages.

Every room had its treasures, an Aladdin's cave of wonder. I met geneticists, anatomists, hair morphologists, physicists and firearms experts and saw rooms full of the most up-to-date analytical equipment. There were plenty of examples of luxury goods confiscated at airports around the world: crocodile handbags, elaborate hats decorated with the plumes of egret and macaw, eagle-feathered headdresses and even a purse made from the bloated body of a toad, with a zip down its tummy. Even more fascinating for me were the lab's reference collections of birds, animals and animal parts, which were used to help the lab identify the material sent there by federal agents. A narwhal tusk, eagles and falcons and all manner of other birds occupied every flat surface. Best of all, hidden from view in chests of shallow drawers were the skins of parrots, bee-eaters and rollers whose glorious feathers of iridescent blues, greens and reds had tempted smugglers to break the law. Next to them, another chest of drawers held the claws of every kind of bear and monkey and the desiccated heads of fox, panther and snow leopard.

One of the laboratory's major functions is to identify the species of confiscated animal parts, and for this they need, and use, their reference collections on a daily basis. These identities are required in order to decide if the seized items are from species on the CITES (Convention on the International Trade in Endangered Species 1973) list and therefore whether any laws had been broken. A lot of the identification work at Ashland is done by members of the morphology team who go by the look of the animal or bird. When a visual identification is impossible, as it might be in the case of a fur coat for instance, the hair morphologists take over and scrutinise the hair shafts under a powerful microscope. If that examination still gives an unclear result – which can happen even with the most expert of eyes, as I was to discover – then the genetics lab gets involved using techniques rather similar to the ones I was planning to use myself.

Another of the lab's investigations required an accurate identification not of a whole animal, nor even a hair, but of a fish egg. It concerned the astronomically expensive delicacy caviar, made from the salt-cured eggs of the female sturgeon. Caviar comes in different forms: osetra, sevruga and, most prized of all, beluga from sturgeon living in the Caspian Sea. Overfishing and pollution have dramatically reduced stocks in the Caspian and in 2005 the US Fish and Wildlife Service banned the import of beluga caviar. Stocks have improved since then, and in 2010 beluga caviar was once again exported from Russia but with a strict annual limit of three tons, compared to higher limits for the cheaper sevruga (seventeen tons) and osetra (twenty-seven tons) varieties. Given the premium on beluga caviar, it is no surprise that unscrupulous dealers have tried to evade the quotas by labelling beluga caviar intended for export as one of the cheaper varieties then relabelling it as beluga before it reaches the shops. A suspected shipment worth several million pounds was seized and samples sent to Ken's lab for identification. Not accustomed to a life of luxury in rural Oregon, neither Ken nor any of his staff was able to give a positive identification based on the effect on the palate. Instead he handed a test sample over to the genetics lab, which ran a species ID and discovered that this was indeed caviar from the beluga sturgeon. The Russian mafia who controlled the trade have never forgiven him, he said with a wry smile.

There wasn't time to see much more, but there was enough to convince me that I must try to arrange a return visit. Although I admired the cornucopia of analytical equipment and the evident expertise of the staff, what was of equal value to me was the director Ken Goddard's attitude to my project. Although he normally steers well clear of Bigfoot and Bigfootologists, he was persuaded, without much difficulty, that my project was different from what had gone before. I had not set out to find Bigfoot, but to examine in a systematic fashion the attributed samples collected by others. It was a great comfort to know that, along with Michel Sartori in Lausanne, here was another professional scientist that didn't think the project was crazy and was prepared to help. As the lab was funded to investigate wildlife crime, we soon came to talk about whether or not killing a Bigfoot was a crime. Apparently not, so Justin Smeja's fears of a homicide charge after he killed what he thought was a juvenile were groundless. Unless he had shot it in Skamania County, Washington State, I added, where an old ordinance of 1969 made this a crime. Ken's eyes twinkled when I told him this.

Finally, before I left I asked about the rumour that I had heard from Greg that there were already Bigfoot samples at the lab.

'Certainly not,' he replied.

'Not even in the secret freezer in the basement? Surely all government laboratories have one of those?' I enquired.

'Not even there.'

I was invited back to Ashland and made three more visits,

spending several days going through my collection of hairs and other material with the lab's experts. The first to undergo scrutiny were the hairs on the 'The Steak' sample that Justin Smeja had given me when I met him in San Francisco. Cookie Smith is Ashland's principal hair morphologist, having taken over from the legendary Bonnie Yates a few years earlier. As soon as we put 'The Steak' hairs under the microscope we could see the small cubic crystals of salt that Justin had used to preserve the sample. The hairs were very short, too short for a bear I thought, and much too pale, although American black bears do occur in a wide range of coat colours. When we followed the hairs to their tips it was clear that they had been cut. Instead of tapering off, the hairs ended abruptly with some separation visible between outer layer, the cuticle and the cortex and medulla. Either this had been deliberate or, since 'The Steak' had been dug up with a spade, it was this excavation which had done the damage. Cookie thought these might be bear hairs, but couldn't really decide.

That same day I took Justin's boot over to Ed Espinoza's laboratory. Ed is the deputy director of the Ashland lab and a man of great ingenuity. He was at work on a very neat mass spectrometer, an instrument that separates the components in vapour given off when a sample is heated up. A fabulous, heavy perfume filled my nostrils as soon as I entered Ed's lab. He explained that he had been at work on some samples of agarwood, a scented wood that has been used for centuries in the production of incense. Its scarcity mixed with a high demand, especially in the Middle East, have made it one of the most expensive woods in the world and stimulated a lively market in contraband. Ed's spectrometer identifies the volatile components given off when a minute sample of wood is heated, and as each wood has a characteristic profile of these volatiles, the species can be identified without much difficulty.

With the voluptuous scent of agarwood still wafting through the lab, I unwrapped the sliver of leather I had taken from Justin's boot and put it on the lab bench. One of the uses of the mass spectrometer is to identify bloodstains from hunters' clothing. To blend in with their surroundings in every sense, hunters don't wash their clothes much, if at all, during the hunting season. The spots of blood that have spattered onto the clothes from animals they've shot often remain in place for a long time. If a Fish and Wildlife agent suspects a hunter of shooting a protected species, like a mountain lion, or killing more bears than his licence allows, then the clothing can be requisitioned and the blood spots examined by Ed's mass spectrometer. In this case it is the volatile elements of the animals' red blood cell pigment haemoglobin that are profiled in the machine. That was an exciting prospect indeed for the sliver of Justin's boot.

The blood on the boot was what I call a first order sample, one that has come directly from an animal so its attribution is not in doubt. This distinguishes it from second and third order samples where the attribution is less secure. A second order sample might come from an object which the creature had been seen to rub itself against, like Dan and Garland's bait tree. A third order sample might be a hair retrieved from a bush close to the site of an 'encounter' but without any direct evidence that the encountered creature was the same one that left the hair on the bush. The blood on Justin's boot was definitely first order, so if Ed's instrument could identify it we would be pretty sure it came from the Bigfoot he shot. 'The Steak', on the other hand, having been retrieved from the scene several weeks later, might easily have belonged to a different animal altogether.

Before vaporising the blood on the boot and injecting it into the mass spectrometer, Ed wanted to have a careful look at the sliver under another instrument to be sure that the clearly visible dark spots really were blood. If not it would be a waste of time and money to carry on with the costly business of firing up the mass spectrometer. We went into an adjoining room where on the bench stood an anonymous grey box about the size of an old-style television. This was a VSC 6000 spectral comparator, the latest version of one of the mainstays of forensic analysis. The VSC 6000 works by shining a narrow wavelength beam of visible infrared or ultraviolet light onto a specimen and then detecting the wavelengths of light that are transmitted back. This technique is frequently used to detect forged banknotes, altered signature on cheques and other similar evasions. It also detects blood. By choosing a setting that illuminates the specimen with infrared light and then scanning across the wavelengths of the emitted light, the instrument picks up a signal from the iron-containing haem molecule which is absolutely characteristic of all blood samples.

Ed opened the instrument and put the sliver onto the tray inside. He started the scan, and as this began a sequence of images appeared on the video screen next to the machine. Each one was an image of the sliver, illuminated by infrared light but filtered to show only a small range of wavelengths. After each scan, the filter changed automatically and the next image, at a higher emission wavelength, began to form. The outline of the sliver was clear to begin with, but as the emitted light filter increased in wavelength it became harder and harder to make out. In the first images I could clearly see the dark spots that Justin had pointed out were the ones where the creature's blood had dropped onto his boot. At around 430nm nothing was visible.

'That isn't blood,' said Ed decisively. If the spots had been blood we would have seen a bright spot, or spots, when the haem fluoresced under the infrared beam. There was nothing. We continued to the end of the wavelength sequence. Still nothing appeared. I asked Ed if he could tell what was in the spot. He said that he could not. I asked if perhaps the haem on the boot had deteriorated to the point where it no longer fluoresced. 'I doubt it very much,' was Ed's succinct reply. 'At this point, if we began by suspecting this to have been blood, we would discontinue the investigation. It is not blood.' No point, then, in firing up the mass spectrometer. Whatever was on Justin's boot was not blood, either from a Bigfoot or from anything else.

While I was still contemplating the implications of this result for Justin Smeja's story, Ed introduced me to another of his instruments. This one was a Fourier-Transform Spectrograph and I wanted to use it to examine the two 'hair' samples that I had received from donors that I suspected from their appearance might not be hair at all. The FTS measures the spectrum of radiation given off by a sample when it is excited by a magnetic pulse. Rather like the VSC 6000 spectral comparator, components in sample emit radiation of a characteristic wavelength that can be tracked and displayed as a series of peaks on a trace. An experienced operator will recognise the most familiar peaks, but the machine is also capable of comparing the emission spectra with a reference collection and coming up with the best matches.

Ed normally uses this instrument to examine fibres of clothing to see what they are made of, so it is perfect for hair and for the questionable fibres whose authenticity I wanted to check. We started with the first sample, which I suspected from its branching structure was a plant root of some kind. Ed spread a single fibre across the small aperture through which the spectrum would be measured, switched on the magnetic flux and the scan started. It was much quicker than the VSC 6000 and in less than a minute we had a good spectrum on the screen. Ed didn't need any time to think. He immediately recognised the clear signal of cellulose in the spectrum. Nevertheless he requested a comparison from the instrument and that confirmed his initial opinion. This fibre contained cellulose, a component characteristic of plants but not found in animals.

Before we moved on to the second suspicious sample, Ed thought we should check what genuine hair looked like. I plucked one from my head and laid it across the aperture of the instrument. Shortly after a very different set of peaks appeared on the screen. This spectrum was nothing like cellulose. Far to the right, a large peak dominated all the others. 'That is keratin,' Ed announced quietly. The peak on the spectrum came from the peptide bond, found in all proteins but giving this particular profile in keratin, the principal component of mammalian hair. Again Ed asked for a comparison with the reference spectra. The closest was with merino sheep wool, also made of keratin.

We turned our attention to the third sample, which I had

thought might be some sort of glass fibre, perhaps from an ethernet cable or insulation material. It shimmered, quite unlike hair, and I was surprised that the donor had still identified it as hair. In fact that seemed so unlikely that I thought it may have been a deliberate hoax, just to see if I identified it as an animal and by so doing discrediting the entire project. As the magnetic probe was lowered onto the hair there was a faint yet audible cracking sound, absent from genuine hair. We didn't have to wait long for the answer. The spectrum appeared on the screen. Ed didn't recognise it. There were peaks but they did not correspond to any peptide bond. This was not a hair, as I had suspected all along. When Ed asked the instrument to check the spectrum against the reference collection, the closest match was silicon dioxide – the main component of glass. I checked the sample number against my database to identify the hoaxer. When I saw which sample this was, I knew at once how it may have been a genuine mistake, in ways I will explain later, rather than a deliberate attempt to fool me.

I spent the rest of my time at Ashland going through the different, genuine hair samples in my collection and seeing which looked like possible primate hairs and which were clearly not. It certainly was no easy task. Primate hairs tend to be quite fine and straight and without the thicker guard hairs typical of many other animals, but there are no outstanding features of primate hairs that make a positive identification easy. Even Cookie Smith was unsure about some of the samples, though she was prepared to give a guarded opinion on others. The easiest to differentiate were the deer hairs, which I cut from the Lab's reference collection of skins. They had a very cellular medulla. The recently retired Bonnie Yates, one of the world's greatest authorities on hair identification, was kind enough to come back to the lab and help me go through my burgeoning collection. Like Cookie Smith, she was prepared to have a stab at some identifications, but by no means all of them. It was Bonnie who told me that individual variation was so great that to give a positive ID from hair you needed many hairs from different parts of the body. She was also the one who pointed out that when she had been unable to positively identify a hair sample, the Bigfootologist would twist this to imply that it was from an unidentified animal. This was such a frequent and irritating corruption of her opinion that she stopped trying to help Bigfootologists many years ago.

On the eve of my departure from Ashland I went for a walk with my wife in Lithia Park in the centre of town. The park was in a blaze of colour from the pink and white dogwood trees. I sat on a bench while Ulla walked on a bit. I was facing a thinly wooded hillside and enjoying the warm spring afternoon. Then about halfway up the hill a mountain lion walked right across my field of view. I wasn't mistaken. It walked like a leopard, with a long tail held in a graceful curve. I didn't even think to myself: 'What on earth was that?' It was a mountain lion and that was that. I mentioned this to Ken that evening when he and his wife joined us for dinner. I knew sightings were uncommon, but neither he nor his wife had seen a mountain lion in the thirty years they had lived in Ashland. Now I knew what it felt like to have seen a Bigfoot. Ulla thought I must have seen a large domestic cat. Ken was too polite to say what he thought. But I know what I saw. It wasn't a cat, or a badger or a raccoon or any of the other suggestions. It was a mountain lion.

We left Ashland and travelled north to meet up with Lori Simmons and her fiancé Adam Davies. We were going to have another go at tempting the Big Guy out from beneath its tree. When I told Ken about this intended adventure he began to get rather concerned. He didn't really believe there was a sasquatch living under the tree, but he thought there might well be a bear. I had never seen a live black bear and he thought I really should, just in case the Big Guy decided he was fed up with being taunted and came out to attack. To rectify this gap in my experience, Ken organised a trip to a nearby wildlife centre where there were bears, both black and grizzly, along with other native animals, including mountain lions.

Wildlife Images near Grants Pass, Oregon is primarily a rehabilitation and education centre run by Dave Siddon, who showed us round. Although the idea behind the centre is to care for injured or abandoned animals and birds before releasing them back into the wild, some residents become acclimatised to humans and cannot be set free. Instead they might land a career in films. We saw the peregrine falcon that had, apparently, starred alongside Tom Cruise. We saw two magnificent bald eagles with equally glitzy film careers. We met 'Tundra', a sad-eyed and very shy timber wolf who was being walked on a lead and had yet to have her first audition.

Further into the park we came across the mountain lion enclosure. The animal behind the fence was huge, much bigger than the one I had seen in Lithia Park. Its body was almost the size of an African lion though its head looked disproportionally small. Even so it wasn't hard to imagine how an animal like this could have killed the jogger without any difficulty. In the adjoining pen were eight black bears just lying around. I didn't like the look of them at all. And they were pretty big too. They looked really mean, with a facial expression that reminded me of a Rottweiler. I had been reassured by the woodsmen I had met that black bears were not a threat and you just needed to keep your eye on them. They were far more frightened of you and would always walk away. That was the theory.

Next to the black bears was an enclosure with nothing in it. Or so it seemed at first. Dave approached the wire fence and began to call out for 'Grizz'. As we waited, he explained that Grizz had been rescued as a cub from Alaska when his mother was shot. Dave had agreed to take him on condition that he would be returned to the wild in Alaska as soon as he had grown. The trouble was that Grizz became far too tame and it was obvious that he could not be safely released back into the wild. He had lost any fear of humans and was very likely to approach anyone he came across, with severe consequences. It was not so much that he might attack a human, though that is always possible with a grizzly, but that he would be shot as a danger to the public. So Grizz was going to live out the rest of his life in Oregon. Less free than in his native Alaska perhaps, but much safer for him, and for Alaska.

Without a sound Grizz appeared, ambling silently towards the fence. To say he was enormous is an understatement. When he got to the fence he raised himself up onto his hind legs. He must have been at least ten feet high. Dave threw him an apple, which rolled back to rest close to the electric wire that was our real protection. Grizz could have demolished the chain-link fence with one swipe of his mighty claws. He sat in his haunches and, with immense delicacy, used one claw to remove the apple to a safe distance from the wire and then crushed it in his jaws. The odd thing was that although Grizz was double the size of the black bears, he didn't look as threatening. His expression was quite mild in comparison. But this is a dangerous illusion. Grizzlies are far more dangerous than black bears.

I was glad to hear that there weren't supposed to be any grizzlies where I was going. Even so I bought an extra can of bear spray, just in case.

25

Knock Three Times

Every day following the extraordinary experience with Lori and the Big Guy I asked myself what could have made that knocking sound coming from under the tree. It could not have been a hallucination, as others heard it too. It must have been a large animal living under the tree. What other explanation could there possibly be? If it really were a sasquatch then this was a unique opportunity: an opportunity to locate one, set hair traps and perhaps even film or photograph the creature at a specific location rather than relying on a rare chance encounter in thousands of square miles of forest. I arranged to return in May 2013 after giving a genealogy lecture in Boston. Adam Davies, who had met Lori the previous year and was now engaged to her, would fly over from the UK and join us. Adam, you will recall, is the British cryptozoologist who has done so much work on the Sumatran *orang-pendek*. He and Lori had camped close to the Big Guy's tree earlier in the year and had obtained a brief trail-cam video sequence of a large bipedal creature standing over their sleeping bags, an experience neither of them will ever forget, though they were asleep at the time. It could have been the Big Guy. They certainly thought so. Adam had shown me the video clip. It was very short, only a couple of seconds, but when he took me to the location I could see from the scale of other objects in the frame, which included a fixed picnic table, that this was a creature at least four feet high with a well-defined dorsal musculature. Was it a bear? It didn't look like it.

We met up at Marblemount and booked into the local inn. What a change there was from my previous visit in March. The snow had all gone and the gardens were full of rhododendrons of different colours mixed in with pink and white dogwood blossoms. The weather was warm, but Marblemount was empty. Even though this was perfect hiking weather, the season here only gets going in June. We were the only ones staying at the inn, and we soon found out that the two restaurants in town only opened at the weekends, which this was not. We managed to get something to eat at the gas station, then prepared for the next day's anticipated rendezvous with the Big Guy.

Though I had packed my bear spray I still felt vulnerable, so bought a fearsome hunting knife at the gas station. I am sure it would not have been much protection, but it made me feel better. I thought I had better report in to the local Ranger's Office and get permission to put up hair traps. This involved a drive back down the Skagit Valley to the National Forest Service HQ at Sedro-Woolley. Fortunately the lady at the reception desk had read *The Seven Daughters of Eve*, which made my eccentric request for a permit rather easier than it might otherwise have been. In no time, the relevant higher authority promised to produce a letter of permission and fax it through the following day. I would then have the authority to set hair traps for a sasquatch. Was this the first official endorsement of such a request, I wondered.

Back at Marblemount, Lori and Adam had returned from their first visit to the Big Guy. He was still there, which was a great relief to me, and was knocking from his lair under the tree as usual. Over lunch Lori explained that she had just told the Big Guy that she and Adam were soon to be married. He had taken the news well and not worked himself into a frenzy. When we arrived back at the tree he was thumping away, louder and more often than on my last visit in March. Lori had laid out the usual offering of green apples and also some Hershey bars, which were another of the Big Guy's favourites. Meanwhile I set up about a dozen hair traps consisting of Gorilla Tape mounted sticky side out on foam pipe-insulation tubes, with the whole construction threaded onto garden stakes and secured with some moss. I had invented this version of a hair-sampling kit at Ashland, having failed to find a manufactured alternative. All I could find on the Internet was an account by researchers in Canada who had strung up barbed wire around a carcass to catch grizzly hairs. This sounded far too brutal for the Big Guy.

I also set up a camcorder and microphone to catch any sight or sound of the Big Guy while we were away. Lastly, using gloves, I put some apples in glass jars which I had wiped with an alcohol swab to remove my own fingerprints. The idea was that if the Big Guy picked any of these up he would leave prints which I could take back to Ashland and have professionally developed and analysed. Only primates have finger ridges, I was told. The thumping was getting louder and Lori was getting nervous so we called it a day and returned to Marblemount.

The following day began quietly enough as we drove to the site. I inspected the hair traps and, sure enough, several of them had short dark hairs stuck to the extremely tenacious Gorilla Tape. I removed the traps and replaced them with fresh ones. The apples and the chocolate had gone so Lori laid out a new selection. Both Adam and Lori pointed out that there were no signs that the apples had been gnawed before they disappeared. The implication was that the apples had been picked up and eaten whole, something a bear was apparently unable to do. I didn't say that I had seen big 'Grizz' do exactly that at Wildlife Images a couple of days before. The Big Guy was rather restrained that morning, despite Lori's attempts at starting a conversation.

It was another beautiful day and we headed back to Marblemount for lunch in the sunshine. We would meet back at the tree after I had collected my permission letter from Sedro-Woolley. It was all going very well. I had some recordings of the Big Guy's thumpings and some hairs to analyse. But the relaxed atmosphere was not to last long.

When I drove back towards Marble Creek and the Big Guy's lair, Lori and Adam were waiting for me on the roadside. Lori looked terrified and Adam explained that the Big Guy had gone berserk. The thumping had grown in intensity and he was roaring and sounding extremely angry. Lori was genuinely in fear of her life and was expecting the Big Guy to come out from his lair and kill her. The reason was obvious, she said. Lori had just explained to the Big Guy that she would be moving to England and would not be back for quite a while. This fresh news enraged the Big Guy and precipitated his terrifying tantrum. Lori could not be persuaded to return to the tree, so we all retreated to Marblemount. After we had settled Lori, Adam and I headed back to the Big Guy. If this was to be the last visit then I wanted to retrieve my hair traps and recording equipment. The roaring had stopped when we arrived, and I collected the traps, camcorder and microphone as quickly as I could. Adam had said that the Big Guy only knocked when Lori was present, but that was no longer the case as we heard a number of knocks. Not loud ones, but he was still at it.

Lori and Adam left the next morning. Lori was still pale from terror and vowed she would never return. She had carried on her father's research for many years, but now felt very threatened by the Big Guy. Not wanting to go back to an enraged sasquatch alone, I followed them down the valley to the Interstate and carried on to stay with Rhett at his home on nearby Whidbey Island while Lori and Adam headed south towards Seattle.

I spent the following day with Rhett on a ferry trip round the San Juan Islands looking for orcas, something I had wanted to do for a long time, though we didn't see any. I set off the next day on the long drive back to San Francisco and home. When I reached the junction with the Interstate that would take me south, I suddenly thought: this is ridiculous. Here I am within an hour's drive of a sasquatch, with all the equipment I needed for a positive identification. I know the creature is there. I may never have the opportunity again. Even if I do return one day, the Big Guy will probably have gone. I just have to go back to Marble Creek. So that is what I did.

The change of plan meant I was going to miss my college's Governing Body meeting. My emailed apologies to the College President, the distinguished biographer Hermione Lee, caught my feeling of apprehension:

Dear Hermione,
    I am writing from the small township of Marblemount in the North Cascade mountains of Washington State where I am pursuing my Bigfoot research programme. For the last few days my colleagues and I have heard a large creature who appears to live under a big fir tree. He, or she, thumps and growls, sometimes very aggressively. I have no idea what it might be. I am going to investigate further.
    The upshot is that I am unable to attend the GB meeting tomorrow and present my apologies. If things here go badly wrong, I hope you will accept my posthumous apologies for future meetings.
    With kind regards
    Bryan

Walking down the track to the Big Guy's tree on my own was far more scary than in company. At Burlington I had bought a 'Go-Pro' video camera, which I mounted on a head-strap. It recorded my progress down the track and would film any creature that appeared. I was certainly frightened and really thought this might be the end of me. At least my final moments would be on film, so long as the 'Go-Pro' survived the Big Guy's lethal attack. All these things went through my mind. I had my hunting knife in my belt and the bear spray in a holster. I test-fired the spray and a jet of foul-smelling yellow liquid shot twenty yards in front of me. Would it stop the Big Guy if he decided to attack, or make him even madder? I really didn't know.

When I reached the tree, there was no knocking to be heard. Maybe the Big Guy has left his lair, I thought. Or maybe he's on his way back. Adrenalin pumping, looking over my shoulder after every step, I did my round of the hair traps. Yesterday's apples and chocolate had all gone, and there were more hairs on the traps. On one of them, round the back of the tree, I could see three long, shiny, golden-brown hairs stuck to the Gorilla Tape. This is it, I thought. After decades of effort by crypto-zoologists, here at last is a genuine sasquatch hair sample. I re-set the traps, opened a new file on the voice recorder and assembled the camcorder. As I drove back to Marblemount, I felt the warm glow of success. I had made the discovery of the century. I had three sasquatch hairs in my pocket and tomorrow, after the Big Guy comes out for the apples, maybe there will be a movie too.

Back in Marblemount, I began to think it would be a good idea to get another witness to the Big Guy's knocking. Not a Bigfoot enthusiast but a neutral opinion. Three old-timers were drinking beer outside the general store, so I went over to speak to them. 'Sure,' they said. 'We know all about the sasquatch in the valley.' They had also known Lori's dad when he was living in the forest. But none of these three seemed particularly good witness material and I left them with their beers. I then thought of the National Park station up the road. I could get a park ranger to witness the Big Guy's thumping; I couldn't do better than that. I drove to the station and launched into my strange request. Very fortunately the ranger on duty, Sage Bohme, had studied human evolution at college so was not as surprised as he might otherwise have been when I explained the scientific purpose of searching for sasquatch. Fortunately, it was nearing the end of his shift. He didn't feel he ought to use company time for this expedition, but in his own he was happy to come with me.

When he clocked off, we set off for the Big Guy's tree, which

Sage immediately identified as a Douglas fir. As is common with Douglas firs, the trunk was divided from just above ground level and two trunks thrust upward to the sky. Before long the Big Guy started knocking again, which was a great relief to me. Sage listened intently. We had both clearly heard the same sound. Three dull knocks coming from beneath the tree. Sage was intrigued, though he said nothing. He started looking around, then climbed down the bluff on the downhill side and made a circuit of the entire tree looking for an entrance to an underground lair. He didn't find one, nor any sign of trampling. When I had pointed out this absence of tracks to Lori and Adam they explained that the lair was reached through a series of tunnels running all through this part of the forest with its entrance probably down by the creek about a hundred yards distant. Sage carried on with his inspection of the site, looking all round the tree and up to the top branches, while I attended to the hair traps. After about five minutes he came over and said in a quiet voice, 'I have an alternative hypothesis,' and directed my gaze up the two parallel trunks.

Fifty feet above the base, but still nowhere near the top, a large side branch from the left trunk had grown over to the other so that they were touching. He got out his binoculars and showed me how the branch was embedded in the trunk so thoroughly that they were in close contact. The ingrowing branch had worn a channel in the trunk, which I could see through binoculars had been polished smooth. Sage's alternative hypothesis was that when the trunks moved in the wind they slid or rather jerked across this tight junction, making the knocking sound. This was relayed down the tree and amplified by the hollow trunk near the base. The sounds appeared to be coming from under the tree, but they actually originated fifty feet higher up.

Sage thought his hypothesis could also explain the changes as the day progressed. At that latitude and in otherwise calm periods, the desert to the east of the Cascades heats up as the day wears on. As it does so it draws air from the ocean inland across the mountains. In the mornings there is very little wind. This was

Sage's explanation for the lack of knocking early in the day, which Lori put down to the Big Guy being asleep. She had noticed that the knocking grew louder and louder throughout the late morning and into the early afternoon. Sage's explanation was that as the wind increased through the day, though barely noticeable at ground level, it moved the tops of the Douglas fir and drew the branch across the trunk lower down. The more wind, the more frequent and the louder the Big Guy's knocking became. When it reached a certain speed the embedded branch slid continuously against the opposite trunk and the growling began.

Sage's hypothesis was testable. If the branch were sawn off, the knocking should stop. I am sure we could have arranged that, but as with many aspects of the project, I needed to avoid going off at a tangent. If Lori wants to silence the Big Guy once and for all, she knows what she has to do.

As I drove away down to Oregon, after thanking Sage for his brilliant analysis, I reflected on my own reaction to the Big Guy. From the moment I heard the first knocking in March, I had become more and more convinced that there was a large animal under the tree. Lori's explanation for the quiet mornings and noisy afternoons seemed entirely reasonable. I even began to believe that the Big Guy was jealous of Adam and wanted Lori for himself. I was rapidly losing the scientific detachment I thought I had and was well on the way to becoming a true believer myself. Thank goodness I asked Sage for a second opinion.

Back in Ashland, when I looked at the short hairs caught on the traps set near the apples and the Hershey bars, they were clearly from deer, with their characteristically frothy medullas. And what of the three glossy sasquatch hairs which would at last identify the sasquatch and set me on the road for the Nobel Prize? As I unwrapped the covering from the Gorilla Tape, the three short hairs became a single long one, dark blond and shiny. Just like Lori's, in fact.

26

The Russians

As we've seen earlier in the book, when it comes to yetis, the Russians do things differently. There is structure to their investigations. Ever since the Snowman Commission was established in 1958, 'hominology', the Russian term for anomalous primate research, has been a recognised scientific discipline. Although the Commission enjoyed only a very brief life before being dismantled by the Soviet government, its enduring legacy is that hominology has never had to struggle for intellectual acceptance as it has done in the West. Astonishingly, three of the original investigators are still active, meeting once a month in Moscow's Darwin Museum just as they have for the past forty years. It was my good fortune that these eminent gentlemen had responded to the press announcement of the Oxford-Lausanne project and were among the first to send me hair samples for DNA analysis. I travelled to Moscow to meet them and hear more about their research in general and the samples in particular. Fortunately, all three speak good English, though I did arrange to have a translator present just in case.

After I gave a short seminar on my project to their monthly meeting, I was excited to sit down with the trio: Igor Burtsev, Dmitri Bayanov and Michael Trachtengerts, who between them have amassed well over hundred years of cryptozoological research. Each has written books on hominology, copies of which they kindly presented to me to take home. I have had to edit the interviews a little. Despite their good command of English, which I had no trouble following as the spoken word, the language difference makes a verbatim account too disjointed. I began by asking Michael Trachtengerts about how he first became interested in yetis. He was quick to correct me, with twinkling eyes and a broad smile that never left his face.

'Not yetis, *almastys*. That is what we call these creatures in Russia. I liked to read about nature and about wild people but what really started me off was when the Patterson-Gimlin film was shown in Moscow in 1975. I lived then near the Polytechnical Museum and saw a notice about this film, so I bought a ticket. When I saw the film, I had at once the sense it was not fake. That is when I wondered if the tales and stories about similar creatures from all over the USSR might have a real foundation. I liked to spend much of my spare time travelling, usually by canoe going for a month or so to the remote rivers of Siberia or in the eastern part of Europe. I heard very interesting stories that seemed to be incredible at first. I started to write up these stories and collect them to publish as books, but after a while I began to try to look for more evidence. I bought a large camera with a 500mm telephoto lens and taught myself how to make footprint casts. I wasn't very lucky. Once I joined an expedition to the Pamir Mountains. We looked around, and I was carrying my heavy camera, but we saw nothing and made camp. An hour later another group came back and said they had seen an *almasty* not far away, and it was jumping from one rock to another.'

I asked Michael whether he had ever seen an *almasty* during his almost forty years of research.

'No, I am unhappy to say. No, I see only footsteps. I see *almastys* only in the tales people tell me, that they saw such creatures and even held them in their own hands. I have spoken to at least a hundred people who have seen *almasty*, mostly in the Caucasus or the Pamirs, but sometimes in other regions. *Almastys* live everywhere, but they are rare. But maybe I am just unlucky. I spent a lot of time in the woods but I have never even seen a bear, even though once my friends saw one just behind me. I saw a lot of footsteps of bears in my village, where I have a summer house, and a lot of footsteps of bears every time I go to forest for mushrooms, for berries and so on. But I have never seen one. But it means that just because I have not seen an *almasty* that they do not exist. I know bears exist of course, but I have never seen one.'

I asked Michael to tell me more about the four hair samples he had kindly donated to the project. The first thing I found out was that he had not collected them himself, which always extends what forensics experts call 'the chain of custody'. Michael was aware of this shortcoming, but did know all about the samples and where they had come from.

'Two of the samples I sent you came from the expeditions to the Caucasus by Marie-Jeanne Koffman in the 1970s. They were caught in bushes close to *almasty* footprints. The third sample also came from the Caucasus, but was collected later, in 1994.

'The final sample came from a very distant place, from Kargapol in the Archangel district in the far northwest of Russia and it has a very interesting history. It was in January in a severe winter. There were military barracks there and some creatures were spotted around this place by a hunter. He saw footsteps of such creatures in the snow and was quite afraid to go into the woods. He followed the tracks until they ended by the wall of the soldiers' barracks. So he thought that the *almastys* might have gone onto the roof and even into the loft to escape the severe frost.

'This was true because later that night two of the creatures, an adult and a child, went inside the barracks, perhaps for the warmth, perhaps to drink from the soldiers' drinking vessels. It was midnight. The older creature just was sitting, waving his hands in front of him to make the soldiers keep quiet. And one of the soldiers, the strongest perhaps, tried to take him out but when he went up to the creature he fainted through fright or something else, maybe the smell, who knows. Then the creature with the child went out of barracks and crossed the parade ground and disappeared. There were about thirty soldiers who saw the creatures and they describe them in quite a lot of detail. The larger one was about eight feet tall and the child was about three feet. They were a brown-greyish colour, and they were not afraid at all. I suppose that they had been to the soldiers' camp several times before because they were familiar with what was inside. The soldiers were very afraid because they never saw such creatures before. But they were from Central Asia and they knew about similar creatures in their native land, so they understand what it was they saw. The soldiers found a bunch of hair on the chair where the grown-up *almasty* had been sitting, and this is what I sent to you.'

This sounded to me like a very tall story indeed, reminiscent of the Edwards Airforce Base yarn. Why, I asked myself, hadn't one of the soldiers reached for his gun? After all, they are said to sleep with them by their side. But I didn't say anything. That was one of the delights of this project. My opinion of what I was told didn't matter a fig. I could be told the most outrageous nonsense and even smile while I listened, then just send the samples to the lab and wait for the truth to emerge.

The Archangel sample was unusual as it was the only one to have come from the Far North where *almasty* sightings are rare. These creatures are far more frequently encountered in the vast taiga forests of Siberia and here they are different from those found further west and in the Caucasus. Along the Ob River in western Siberia for example there are dozens of eyewitness reports, and according to Trachtengerts these Siberian *almasty* are dangerous. The woods are their domain and they don't like people encroaching on their territory. Michael told me that they want to eat people because they need meat. They can kill any creature quite easily. They are not afraid of dogs, but dogs are terrified of them. Even trained hunting dogs bred to be fearless and aggressive will run from an *almasty*. The aggressive tendencies of the Siberian *almasty* towards humans contrasts with the peaceful nature of the North American sasquatch, for example. But the fear they instil in dogs is something I heard many times in my travels in the United States.

A lot of *almasty* reports came from the Caucasus, an area which I became extremely interested in when I investigated the case of Zana, the wildwoman, as you will hear. These reports of sightings have been thoroughly documented by Russian scientists since the Snowman Commission was established. Boris Porchnev and his protégé Igor Burtsev had both worked there, but it was a remarkable woman Marie-Jeanne Koffman who spent the longest time researching the *almastys* of the Caucasus. Koffman was born in France and moved to Russia in 1935. She was a medical doctor and also became a celebrated climber and, most notably, commanded a battalion of Russian alpine troops in the Second World War, for which she received several Soviet military decorations.

I asked Trachtengerts what he knew about the Caucasus region and its *almasty* inhabitants. Which turned out to be rather a lot.

'In the Caucasus there are some huge creatures. For instance, we have a story about when a big man, more than six foot tall, met such a creature in the mist, which was twice as tall as he was. These are very hairy creatures and their colour can vary from nearly white to almost black. The hair is about four inches long except on the head where it is longer, like a woman's. They are fond of water and like to bathe. And they use tools too. They try to steal combs. In the villages, if you leave a comb somewhere outside your house, they will steal it and use it to comb their hair.

'They eat mostly vegetables and also some meat, although it is not easy to find meat in Caucasus. But usually, they eat vegetables – and mice. In the Pamirs, across the Caspian Sea, the *almastys* are white, but in the Caucasus, under their hair, the skin is black. Some people say they have African features.'

Both Dr Koffman, and Dr Porchnev thought these *almastys* might be some form of surviving Neanderthal.

When I asked about the current state of affairs in Russian hominology research, Trachtengerts' normally beaming smile left him for a moment. He felt, in common with many of the other enthusiasts I encountered, that scientists didn't take proper notice of what they did. They did not want to investigate these creatures. These days he felt reluctant to speak about these creatures for fear of ridicule or isolation. Nevertheless, the Darwin Institute, a well-respected academic institution, still hosts their monthly meetings, so the trio of scholars have not been entirely excommunicated. This link to a functioning institute, however tenuous it may be, is invaluable, and is the only one of its kind in the world. I did make a weak defence of scientists' lack of interest in the trio's research. There seems somehow to be an expectation that scientists ought to be interested enough to work on these creatures. Speaking as a professional scientist myself, I am quite sure that there would be no shortage of offers, but only once there is some hard evidence to go on. Stories of gigantic creatures sharing a room with a unit of soldiers is nowhere near enough to convince any mainstream scientist to invest the time and effort to make a committed research project. I came away from my conversation with Michael Trachtengerts reflecting with some sadness that forty years of research with little to show but travellers' tales and a few hair samples had reduced the hominolgy seminars to nothing more than a monthly social event. If any professional scientist had spent even three years, let alone forty, with such a meagre output, then he or she would have been strongly encouraged to move on to something else, or risk being fired. But as you will see, I was too hasty in dismissing their apparent lack of progress.

Igor Burtsev, the leader of the Russian scholarly trio, is a few years over seventy. He is a tall, slim man, with a fine, angular face. While Michael Trachtengerts was always smiling, Burtsev wore a much more serious expression. He has been involved with the *almasty* in Russia for forty-eight years, since 1965, and is well known internationally, particularly in North America, where he has been on many visits to study Bigfoot. I came across several references and photographs of a young Igor, often in the company of his mentor Boris Porchnev, when I was reading through the Heuvelmans archive in Lausanne. Burtsev told me that he became hooked on the *almasty* in 1965, when a friend suggested that he should go to the Caucasus one summer to look for 'snowmen'. That year Igor and his wife Alexandra spent their holidays in the Caucasus, where they met up with the expedition led by Marie-Jeanne Koffman, who by then was more or less permanently based there. That is when Igor 'caught the bug'.

'We were helping Jeanne Koffman to repair her car and the house she rented for the expedition when a neighbour came round and said, "Oh, there is a woman in the next village who met an *almasty*. After that she became ill and is now in her house. If you are interested you can speak to her." So we went along in the car we had just repaired and sat down in the woman's house and heard her story. I had no doubt that the story was real, that she really had met an *almasty* in the forest. After that I spent all my spare time looking for the *almasty*. One month every year until I retired, and now even more often.

'Five years after I started I became ill and could not go on expeditions for a while, so I began to help Dr Porchnev with his researches. He was also very interested in the Caucasus and did a lot of work with Dr Koffman. But it was only in my leisure time. I was not paid for these activities. I am a doctor of history, but in another field, not in snowmen. In the last ten years, since I retired, the *almasty* is the main occupation of my life. The main purpose of my life.'

I had to go further so I asked Burtsev more about the creatures he has spent so long studying. I was surprised that, almost at once, he began to stress their paranormal abilities.

'Years ago we thought this was just an animal, like an ape or maybe a Neanderthal. We started to make something to tranquilise them, but it never worked. We came to understand that we were dealing with a very, very intelligent creature and that it had paranormal abilities. They can use telepathy. If they don't like you to come too close, they will just stop you. You cannot overcome this obstacle. You just cannot move. You are walking down a path in the woods, and you just stop in your tracks. Typically a man is going through the forest and begins to feel some discomfort, something that is bothering him. He does not understand what it is. He feels badly and he does not know the reason. He starts to look around and, "Oh behind that tree is a big hairy creature standing there," and after that he understands what was bothering him mentally.'

I asked Igor if this had ever happened to him. 'Many times,' he replied, but like Trachtengerts, he has never actually seen one of these creatures himself. Nonetheless Igor has documented hundreds of eyewitness accounts from people more fortunate than himself.

'A military doctor was one of the first eyewitnesses I interviewed. He found this creature in the Caucasus during the war, World War Two, in 1941. A local detachment of soldiers had captured such a creature, a hairy giant. Maybe it was ill, I don't know. The soldiers brought him into a barn and kept him there in the winter time. They called the military doctor to determine what sort of creature it was. Maybe it is a deserter from the army or maybe it is a spy? The military doctor asked, "Why you keep him in the barn? This is winter. It is cold. He is without clothes, just naked." They replied, "We brought him into the warm but he start to smell badly. He was sweating and that is why we brought him into the barn." The doctor started to examine the creature and saw lice on him that were much bigger than humans normally have. On the brow and on the side of the face, he saw so many lice.

'He didn't know what sort of creature this was. It wasn't a deserter or a spy and they didn't know what to do with him. They asked for orders. Should they kill him or let him go? In the end that is what they did. They let him go.'

I had to ask why nobody had taken a photograph. Igor could

only say, rather feebly I thought, that they did not think of it or that they didn't have a camera. But what about during the hundreds or thousands of other encounters in those days? And now that almost everybody has a mobile phone, surely there should be lots of photographs?

'We must bear in mind the resistance of the creatures themselves. They don't want to be photographed, they don't want to be registered.'

This was getting silly. All it would take to convince the world that these creatures exist is some really good unadulterated photographs. How can the creature possibly know it is going to be photographed? And if *almasty* and Bigfoot somehow use their telepathic powers to avoid being photographed, how can you believe, as you have said you do, the Patterson-Gimlin film or, even more recently, the mobile phone pictures from Siberia? I was referring to the widely reported video of a yeti taken by 12-year-old Yevgeny Anisimov in Kemerovo, Siberia. Igor Burtsev is quoted as saying says he believes the footage is genuine.[1]

Igor evaded these questions altogether, preferring instead to launch into the familiar diatribe about how the Patterson-Gimlin film was labelled a fake without the benefit of the serious study he has given it. Could there have been a force field around Bluff Creek, where the film was shot, that neutralised the Bigfoot's telepathic powers? However, I didn't bring that up, even though I felt sure that Igor would have come up with an answer.

We got on to talking about hair samples and I got the usual stuff about how hairs that Igor had given to specialists had often been reported as unidentified. As we have touched on before, this more often than not means unidentifiable, not unidentified. Then we got on to DNA and Igor told me that he had sent some samples to an American lab, but they could only sequence mitochondrial DNA, which Igor thought was not enough for an identification. But, as I hope you will realise by now, mitochondrial DNA is in fact the very best way of identifying the species origin of a hair sample. To support his view that mainstream science journals consistently refuse to publish the results of DNA studies, he cited Dr Melba Ketchum's work in which she studied over one hundred Bigfoot samples from the US and Canada and concluded that the sasquatch were hybrids between humans and some other creatures. I have dealt with this study elsewhere but one of the tragedies of the Sasquatch Genome Project is that even well-informed crypto-zoologists like Igor Burtsev cannot see how very flawed the work is. He puts its general rejection down to the prejudices of mainstream scientific journals rather than to its poor quality and unsupported conclusions.

I had to stop at that point in our discussion as I was required to go to another appointment on the other side of Moscow. Fortunately, Marcus Morris was with me and able to take up where I had left off. He had been listening to Igor's ducking and weaving and straightaway challenged him on the possible outcomes of my DNA project. Marcus began by stressing the important ability we had developed to remove all human contami-nation from the surface of a hair sample.

'The important thing about Bryan's work is that it puts an end to all the debate about human contamination. The best thing about his technique is that it completely removes contamination. But what do you hope for from the results?'

'I would be happy if he said this was a positive result,' Igor replied. 'If he proved that this creature is different from human species, maybe like a Neanderthal, or a different kind of hominid. That is what I believe the creature is.'

Marcus then pressed Igor on whether he would believe the results of the DNA tests on the hair samples, whatever they turned out to be.

'I do not know, I'm not a geneticist. I am not aware of Bryan's capabilities. I am not a geneticist, and I am not a specialist in this field. I can only say that if he said "It is human . . ." But that might mean he could not find the difference, which is less than one per cent, between DNA of *Homo sapiens* and this creature.'

(You will by now realise, I hope, that Igor is mistaken on this point.)

Marcus continued: 'But that is what Bryan and I worry about. If he gives you a result that isn't the result that you would like to hear, you will say, "The test is not good enough, do a better test," because you passionately want to believe the *almasty* exist. So much so that if Bryan says, "I'm sorry, but the samples you have given me that you believe are from the creature are actually from a bear or a goat," you will not accept it, and you will be so disappointed that you will find ways of saying, "I don't believe in science."'

'Science cannot distinguish the difference between this species and *Homo sapiens*,' Igor again insisted.

'But it can, that's the point. How can you say that, though, if you admit you are not a geneticist? Bryan is a geneticist and he says he can,' Marcus replied robustly. 'There comes a point where you have to believe the scientists even if the news is bad. The news may be good, who knows. But if it is bad you still need to believe it. And if the results says this is goat or bear, or something else it does not mean that *almastys* do not exist, just that that particular sample was not from a primate.'

After my conversation with Igor, and when I heard what he had said to Marcus, I was beginning to think we were both wasting our time. It looked very much as though Igor would believe the DNA results from the hairs that Michael Trachtengerts had donated to the project only if they showed the result he wanted, but not if they didn't. Actually that is a situation I have faced many times as a geneticist, and it has not always made me popular. But if all DNA did was to give the result you wanted, why bother doing any tests at all? The power of genetics is precisely that it doesn't care what you think. However much you want to get a particular answer, the DNA isn't going to listen. So when it does come up with a result that you were hoping for, you can believe it. Surely even Igor could not believe that the *almasty*'s telepathic mastery would stretch to being able to change the DNA sequence of one of its hairs from Neanderthal to goat? But you never know with a true believer. Hume, I'm sure, had to put up with much worse.

The appointment that took me away from my conversation with Igor Burtsev was with the third member of the Moscow trio, Dmitri Bayanov. Like Trachtengerts and Burtsev, Bayanov has written books on the subject of 'hairy hominids' in Russian, but also in English, without the need for a translator. In many ways he is the most intellectual of the Moscow scholars. Like the others, he has been in it from the beginning, but his motivation to get involved with hominology was one I had not come across before. Whereas many people who I met or read about had their interest triggered by a direct experience, either seeing one of these creatures or coming across footprints while trekking through the wilderness, Dmitri Bayanov had more of a sociological viewpoint. He was born in 1932 and his family moved from Moscow to the relative safety of Tajikistan in the early 1940s as the German army advanced. They nearly starved. While in Tajikistan he heard the common rumours of strange 'wildmen' that were said to inhabit the forests and the mountains. Returning with his family to Moscow after the war, Bayanov eventually went to college to study anthropology, specialising in folklore and mythology. The privations, however, continued. At one stage he was reduced to collecting empty beer bottles from the streets to make ends meet. His main reason for becoming intrigued by the *almasty* and related creatures was more philosophical than biological. He considered that studying these creatures, if they were in some way related to ourselves, might explain the root cause of all the troubles of Russia and the Soviet Union that were all too clear to him. Quite how this would help he did not elaborate.

Bayanov met Boris Porchnev in 1964, read his book and from then on became fascinated by the prospect that, as Porchnev advocated, the *almasty* might be a surviving form of Neanderthal. He joined the *almasty* expeditions to the Caucasus led by Marie-Jeanne Koffman and interviewed dozens of eyewitnesses, whom on the whole he believed were telling the truth. He told me about the long history of 'wildmen' in European and Russian folklore, and that the founder of modern taxonomy, Carl Linnaeus, in his magnum opus *Systemae Naturae* (1758) included a special classification for them. While modern man was dubbed *Homo sapiens* by Linnaeus, literally '*wise man*', he coined the term *Homo troglodytes* meaning '*cave man*' for the wildman. Like his two colleagues, Bayanov has seen footprints but not the *almasty* itself, though over the years he has interviewed plenty of people who have. Like many of his witnesses he believes the *almastys* are very aware of human presence and keep themselves hidden.

Although Dmitri Bayanov gave me very much the same information as the other members of the Moscow three, we have kept up a correspondence and debated the philosophical nature of scientific enquiry, something I had hoped would emerge from this project. As with Burtsev, I was surprised that the intellectual Bayanov thought that only 'positive' results should be made public, as these extracts from our correspondence reveal.

I was trying to track down the whereabouts of Dr Jeanne-Marie Koffman, if she was still alive, to see what she knew about the aboriginal 'forest people' of the Caucasus. Following my DNA investigations on Zana, which I cover in Chapter 29, you will understand why I had become intensely interested in them. Dmitiri told me that Koffman was now well into her nineties, had returned to France and was being cared for in an old people's home.

It was later in the same correspondence that I responded to Dmitri Bayanov's surprising position regarding the genetic testing.

Dear Dmitri,

Many thanks indeed for your email, and the information about Dr Koffmann, sad though it is . . .

I am a little surprised that you feel you can only accept results that are positive for hominology. I hope that getting a proper article published which does at least give a method for unambiguous species identification of hair samples is progress of a sort and that we can put behind us arguments about the accuracy of species identification. That surely is positive for homology . . .

    With best regards, and thank you for your thoughts.

    Bryan

To which I received this reply:

Dear Bryan,

    Let me try to explain once again my mood and position. You and I, along with the rest of the scientific community, know that Socrates existed. We don't need bones or DNA evidence for that. Bones and DNA are not the only criteria of reality. I do know that homins exist without the evidence of bones and DNA. But the scientific community does not know that homins exist without such evidence. If you say in your article that DNA evidence for the existence of homins has not been found, our critics will say: Sure enough it can't be found because the whole thing is a myth. So it's in hominology's interest that you publish positive results of testings, not negative, at least in your initial publication on this subject. It's as simple as that . . .

    Best regards

    Dmitri

Despite this surprisingly retrograde attitude, it turned out that the decades of work by the Russian trio had not been wasted, as we will soon see.

27

The Laboratory Reports

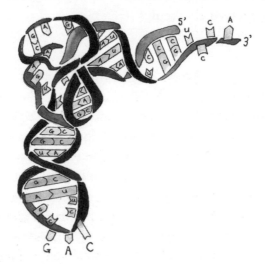

This whole project had been a gamble. I had no idea when I started whether I would get anywhere near enough hair samples to test, nor did I know how many of these would contain enough DNA to make a species identification feasible. Neither, of course, did I know what my DNA tests would reveal. I realise I have strained your patience to the limit, but in this chapter and the two following we will at last go over the detailed results that emerged from the lab and see what they reveal about the mysterious creatures that at once enthral and frustrate all of us.

Over the whole course of the project I received a total of ninety-five hair samples. The majority were generous contributions from personal collections. The others I found by ferreting around in museums around the world and which I was allowed to test thanks to the enlightened generosity of their curators. The sample total was well over the initial target of twenty, which was the minimum I thought I needed to make this a solid project of a standard that could be published in the scientific literature. As I have already indicated, I definitely did not want to do one-off flash-in-the-pan tests on just a few samples.

I whittled the ninety-five hair samples down to thirty-seven through a combination of their microscopic appearance and their provenance. There were many more that I would have loved to have analysed had funds permitted. It was always a difficult decision to put a sample to one side and not to take it through to DNA sequencing. Every time I did so I was troubled by the nagging feeling that maybe I had missed the 'golden hair' that would prove once and for all that a new primate or hominid was stalking the woods.

From a technical point of view, this turned out to be a spectacularly successful project. Even though many of them were small, old and in bad condition thirty of the thirty-seven samples that I sent for analysis produced enough DNA to sequence. More than that, all the sequences were of good enough quality for unambiguous species identification. It was also very clear from the results, as you will see, that the painstaking decontamination steps had succeeded in completely removing any hint of human contamination. I cannot stress enough how important this unglamorous part of the analytical process has been to the project. We have seen how often contamination has ruined previous attempts at genetic analysis in this controversial field.

I had deliberately anonymised the hair samples that went to the lab. All the lab technicians knew was the sample number, nothing else. I had also given instructions that I was not to be told the result of the DNA analysis before the end of the project and certainly not until I had interviewed the donors. All I needed to know for each sample was whether or not the lab had come up with a clear species identification, but not what that identification was. To do otherwise, I felt sure, would influence the way the interview went and bias its outcome. Had I known before talking to a donor that the sample he or she had submitted in the sincere belief that it was from a Bigfoot or an *almasty* had, by DNA analysis, turned out to be from a goat then I would not be able to disguise the fact that I knew. This turned out to be a blessing; when pressed by a donor for an answer, I was able honestly to say that I just didn't know the result.

However I was obliged, reluctantly, to tell some of the donors the results of the analysis on their samples during the filming for the Channel 4 documentary series 'The Bigfoot Files'. I had no desire to reveal results to donors on camera, but one of the many things that makes a film very different from a book is the enthusiasm of film-makers for 'the reveal'. The prospect of eliciting an emotional reaction to a result, especially when immediately followed by the 'So how do you feel about this?' question I found a distasteful prospect. I warned donors that if they agreed to take part in the film they would very likely be asked to receive the news of the results from their sample on camera. Donor participation in the science project did not require them to be part of the documentary, and a few donors declined to be filmed at all. Despite my caution, most people did agree to risk humiliation, and it was agreed that I would be the one to break the news. If anyone was going to do this I thought it should be me rather than the presenter, Mark Evans. After all, it was part of his job as a presenter to wring as much emotion as possible from these potentially sensitive encounters, especially when the news contradicted long-held beliefs about the origin of their samples. In the end we agreed he should take over once I had conveyed to the donors the essence of the genetic results.

From an entirely scientific point of view, it was not necessary for me to dwell on donors' reactions to the results, and I am not going to make a meal of it here. Besides, I knew from experience that people need time to adjust to the results of genetic tests, especially if it was not what they were expecting. On those occasions in *DNA USA*, for example, when I did give volunteers their own DNA results, the immediate reaction to surprising news was rarely dramatic. They needed a day or so to digest it, so it was much more constructive to talk to them some time afterwards when there had been time to reflect.

Of course in this project each and every donor was as keen

as mustard to find out what their own sample revealed, and I had found it hard not to peek at the results myself. When they were all in, I asked my secretary to compile a list of the sample numbers matched to their species identification from the lab. I had the list that told me the source and number of each sample. Both lists went into separate white envelopes. I sat down with my researcher Marcus Morris and Ulla, who had been my *de facto* research assistant on most of the sampling expeditions, to work our way through the lab results.

We were in my study in Oxford overlooking the River Cherwell and the meadows beyond. It was a bright day in early summer and the trees on the banks of the river had just begun to burst into leaf. We each had copies of both lists in their white envelopes and set them down on the table in front of us. We knew that once we opened the envelopes there was no turning back. We would have discarded the 'cloak of ignorance' that had, up to that point, protected us in our dealings with the donors and with the press. Was there any final action we needed to take that would be compromised once we knew the results? We could not think of anything. At last, after more than fifty years of ambiguity, argument and uncertainty, we were about to discover the truth.

Simultaneously, we tore open the envelopes and removed the two lists. There was silence as each of us studied the sheets in front of us. On just two sides of paper was the culmination of tens of thousands of miles of travel, tens of thousands of pounds in analysis costs, months of intense lab work and the hopes of our sample donors and of cryptozoologists the world over. No longer would the only DNA work in cryptozoology be that of scientists whose methodology was suspect or whose commitment was open to question. For the very first time, here was a proper series of results whose accuracy could be entirely relied on.

I had a lingering anxiety that a lot of the samples would be identified as human, much as in Dr Ketchum's series. However, I was pretty sure that Dr Ketchum had been misled by her inexperience in dealing with the curse of contamination and that most or all of her samples that she took to be human were probably misidentifications because of that. I knew that we had paid meticulous attention to decontamination, not a glamorous part of the process perhaps but absolutely essential in this kind of work. So I was not expecting many of our hair samples to return a human sequence. Looking down the list it was a relief to see that only one was identified as coming from a human. We had successfully circumvented the curse of contamination. Although I do realise that this will not excite many people unfamiliar with the agony of doing ancient DNA research, really it was nothing short of a triumph.

The only human hair was sample #25072. I looked up the details. It had come from the window ledge of a cabin in southeast Oklahoma. The donor, a Bigfoot enthusiast, had been in the cabin when it was attacked by a sasquatch. The creature had tried to enter through a bathroom window left slightly ajar. It failed, but the donor noticed a single light-blond hair about eight inches long that had caught on the sill. He was so convinced that he had come under attack by a sasquatch that he had no hesitation in sending the hair to Oxford.

It looked from the DNA result as if the intruder was not a sasquatch at all but a human. However, as the 12S RNA analysis technique could not distinguish between a human and a Neanderthal, I immediately asked the lab to run a second test. This time I wanted to know the HVS1 sequence of the mitochondrial control region, which as we saw in Chapter 13 is very different in *Homo sapiens* and Neanderthals. The lab soon reported that the additional analysis had shown the cabin had not, after all, come under attack by a Neanderthal but by a modern *Homo sapiens*, probably a European, judging by the fine detail of the DNA sequence. I was reminded of the Ape Canyon incident from Chapter 6 where another cabin was attacked by humans mistaken for apemen. As all the other hair samples had returned a non-human sequence, which meant the decontamination steps had worked, I was confident that sample #25072 was genuinely human.

The sample lists were separated into continental regions, and since the largest category were Bigfoot hairs sent from America, this was the first we went through in detail together. Two of the early samples came from Washington State. Sample #25028 was a clump of roughly thirty mid-brown hairs of medium thickness about 8cm long. They were from Enumclaw, King County, Washington. Enumclaw lies on the site of an ancient mudflow from nearby Mount Rainier, with the forested slopes of the volcano not far from the town. The donor was hiking through the area when he heard a large animal moving through the undergrowth accompanied by a low whistling. Seeing the clump of hair caught six feet up in a fir tree close by, he retrieved it and put it in an envelope. When he showed it to naturalist friends, they were unable to link it to any of the local fauna but the lab report showed that sample #25028 was from *Ursus americanus*, the American black bear.

Another early sample donated to the project (#25029) came from further north near the Snoqualmie Pass in the northern Cascades. It was a large clump of very fine hair, red-blond in colour and about four inches long. The donor had seen two sasquatch at the site, most recently in March 2010, and had been back there some thirty times in the two years before he came across this sample in May 2012. Around noon that day, he was trying to get a vocalisation response to his howling when he heard a loud low growling nearby that 'almost bowled me over' as he put it. The clump of hair was lying on the gravel surface of the trail he was following, but it was not from the sasquatch. The lab identified #25029 as a member of the genus *Canis*. The mitochondrial 12S RNA test does not distinguish between different species of this genus, so the hair could have come from a wolf, coyote or a domestic dog.

Sample #25035 was sent to me by Maxwell David whom I met in Willow Creek, northern California. Maxwell, a pseudonym, was an intense, enthusiastic, dark-eyed attorney from Chico in the Central Valley, a hundred miles north of Sacramento. Chico is a stop on the Amtrak Starlight Express, which I rode a lot as I shuttled between San Francisco, Klamath Falls, Oregon, Portland

and Seattle. Maxwell moved to Chico from Arizona, where she had spent twenty-five years of field research in the rugged cactus desert of the Tonto National Forest, during which she had several sasquatch 'experiences'. We had a long email correspondence before finally meeting at Willow Creek. It was then that she explained her experiences of Bigfoot's luminescent eyes, which made them appear to glow in the dark. She concluded from this that sasquatch possessed a *tapidum lucidum*, the reflective layer directly behind the retina that is a feature of many mammals. The *tapidum lucidum* gives rise to the phenomenon of 'eyeshine', when eyes reflect back to an observer with a light source such as a torch or a car headlight. Eyeshine helps animals to see in dim light by giving the retina cells another opportunity to detect the reflected photons. For this reason, eyeshine is particularly well developed in animals that are active at night. Deer, cattle, sheep, cats and some dogs are easily picked out in the dark thanks to eyeshine. Other than some lemurs and tarsiers, primates, at least the known species, do not have a *tapetum lucidum*, so have no eyeshine, though that is not to say that a mainly nocturnal primate could not have evolved this facility.

The sample that Maxwell sent me consisted of four hairs about two inches long with clear dark and light striations at half-inch intervals along their length. She had collected them from a 'gifting site' in the Tonto National Forest, north of Phoenix, Arizona. Maxwell had been visiting the site regularly over many years, slowly acclimatising the sasquatch to her presence by offering them food. At first she left apples, a favourite bait, out in the open. Since any animal could take these, she started to put the apples inside a plastic sandwich box with its lid fastened by a clip so that only a creature with an opposable thumb, like a primate, would be able to get at the apples. And sure enough, the box was unfastened and the apples removed. On one occasion the lid was even replaced and fastened with the clip. Whatever was taking the bait must be capable of considerable manual dexterity. It was after one of these occasions that Maxwell found four hairs inside the opened box, which she sent to me in Oxford.

As with the others, I did not know the result of the analysis when I met Maxwell in Willow Creek. The meeting was in the fabulously untidy 'Bigfoot Bookshop' close to an enormous wooden carving of a sasquatch. The bookshop was presided over by the enigmatic Steven Streufert Jr, not so much an active Bigfoot researcher, more of an observer of Bigfootologists with an eager blog following. Maxwell repeated her assumption that the hair sample must have come from a primate because only a creature as intelligent and dexterous as a monkey, ape or human could have opened the box and clipped it shut. Unfortunately this optimistic assumption was wrong. Sample #25035 came from none of these but from *Procyon lotor*, a raccoon, an animal apparently quite capable of undoing a sandwich box, removing the contents and doing it up again.

From not far away, in Texas, I had been sent a well-known sample (#25023) with its own nickname. 'The Dreadlock' had been found by a rancher on his land wrapped around a thorn bush. The large clump of hair was composed of hundreds of long, thick, red-brown hairs matted together, hence the nickname. The donor recalled seeing giant footprints on his ranch in the vicinity of 'The Dreadlock' and thought it almost certain that the sample he was giving me came from the same sasquatch as left the prints. There were no other candidates, he assured me. The lab report disagreed. Sample #25023 was from *Equus caballus*, a horse.

Another sample from the Southwest (#25167) came from Brenda Lewis, accompanied by a very dramatic provenance. She found it on the Navajo reservation in New Mexico after a disturbing incident. One dark night in June 2012 she heard a huge commotion. Two truckloads of men with guns were driving very fast as if they were chasing someone or something. The following day she and a friend went to investigate. They found a set of large footprints fifteen inches long and five wide moving very fast up a steep hillside, judging by the length and depth of each stride. Brenda and her colleague followed the trail over the brow of the hill where they came across blood, hair, bullet casings and tyre marks. It looked as though the creature, whatever it was, had been encircled by the trucks and shot. Brenda concluded that this creature was probably a Bigfoot. She had seen one in her back field a few years earlier and her neighbours had also had similar experiences, with the added ingredients of growls, yells and whoops. The lab report revealed the identity of the creature that had been pursued by the bloodthirsty Texans. Sample #25167 came from *Ovies aris*, a sheep. The Navajo are prodigious shepherds.

From further north, in Washington State, I received a sample from Portage Island off the Lummi peninsula. Like the Lummi tribal lands on the mainland, Portage Island was reportedly full of sasquatch, with howling and growling regularly reported. The island is connected to the mainland by a causeway that floods at high tide. It is flat, heavily wooded and uninhabited, at least by humans. This sample had been located by an enthusiast who had spent several days camping on the island in the hope of seeing one of the resident sasquatch. After a night disturbed by crashing undergrowth, snapping trees and a lot of loud vocalisations, my donor found a bunch of hairs clinging to a piece of tree bark that he was certain had not been there the previous evening. He sent me one of the hairs for DNA testing. It was about 5cm long, of medium thickness and light brown in colour. This was the last one left in his collection as he had sent the others to Dr Ketchum. But one was enough and the lab found a DNA match for #25030 to *Bos taurus*, the cow. When I told the donor the result, he revealed that Portage Island was home to a herd of feral cows.

Inland from the Lummi peninsula lie the northern Cascade Mountains, from where, near Marble Creek, Lori Simmons' late father had collected a clump of hair he was sure had come from the family of sasquatch, which included the Big Guy he had befriended. He found the hairs lying on an overgrown logging road close to an old gold-mining trail that he and his brother used as a hunting route on that day. Her father had that sixth-sense feeling that something was watching them so he and his brother left in a hurry, followed by the sound of a large animal crashing through the forest in hot pursuit. Over the years when he lived nearby, he got to know the sasquatch family very well and often brought Lori to the area when she was a child. Of course it was Lori who introduced me to the Big Guy and accompanied me on the alarming return visit to his underground lair.

Hair sample #25069 was donated by Lori, though originally found by her father on the logging road fifteen years earlier. It had been kept in an envelope in the interim. Inside were thirty black hairs of medium thickness and about 12cm in length. Given the provenance and the association with the Big Guy, I sent the hair straight to the lab without a preliminary microscopic examination. When the results came back sample #25069 was identified as coming from either *Odocoileus virginianus* (white-tailed deer) or *Odocoileus hemionus* (mule/black-tailed deer). As the latter is found abundantly in the Cascades, and given that the clump of hair was black not white, that is probably what it was.

Sample #25081 reached me from east of the Rockies, from Michigan. Though the majority of Bigfoot/sasquatch sightings have come from the three Pacific states of California, Oregon and Washington, and from British Columbia, there have been sporadic reports of these creatures from almost every state of the Union. So to receive a sample from Michigan was not such a surprise as you might suppose. It was found in March 2012 near Shellenbarger Lake, following a spate of sasquatch sightings. It struck me as an unusual location as the lake is close to a country club and not much further away lies the town of Grayling. Nevertheless my correspondent assured me that the woodlands around Shellenbarger Lake were a well-known sasquatch hotspot, with many reports of close and sometimes alarming encounters. I was sent a clump of about a hundred fine, mid-brown hairs 12–15 cm in length that had been found in the middle of a forest track-way leading to a campsite close to the lake shore. The identity of the Shellenbarger sasquatch sample was revealed by the DNA analysis. Sample #25081 was from *Erethizon dorsatum*, a porcupine.

So far in this chapter I have given the results from what I

call third-order samples. Hairs collected close to a site where a Bigfoot had been seen or otherwise 'encountered' with no certainty that the hair had come from the creature itself. A third-order sample may have come from another animal in the vicinity and been misattributed to Bigfoot. A sample becomes what I classify as second-order when a sasquatch has been observed to come into contact with something, perhaps a tree or a bush, from which the hair is subsequently recovered. Second-order samples are much rarer than third order and the attribution to the sasquatch is much stronger. Whenever I heard about a second-order sample, I made an extra effort to get hold of it. A case in point is #25212, which Peter Byrne told me about in October 2013. His Bigfoot 'buddy' John Cordell had heard about an incident in the 1960s that occurred close to Roseburg in central Oregon. The source was from a local resident, Betty Croft, whose parents' truck had crashed into a Bigfoot in the 1960s leaving a clump of hair with scraps of skin stuck on the trailer. Betty's mother had placed the hair and tissue in a bag and kept it as a souvenir of the mysterious encounter until her recent death. Her daughter Betty had kept all her things, and knowing I was interested, Cordell had discovered from a phone call that she had located the hair and would be very willing to contribute some of it to the project.

I made sure that I passed by Roseburg on my way back from Pacific City, Byrne's home town on the Oregon coast, to San Francisco. Betty met me at the Apple Peddler diner just off the I-5 where, over a light salad and a glass of water, she told me the full story. In 1967, her parents had bought a parcel of land near Sutherlin, a little north of Roseburg. At weekends they would take their Volkswagen and trailer down to the plot to clear it ready for the house construction. One evening they were heading home, with the small trailer on the back carrying the tools they were using on the property. Almost as soon as they turned onto the road, they both saw a Bigfoot cross right in front of them and felt an impact as the trailer swung out and hit it hard. The creature was about seven feet tall, very muscular and hairy with long arms by its side – a classic sasquatch. Her parents had seen very clearly that it had walked across the road on two legs. After the collision, Betty's parents stopped to see if the injured creature was by the roadside, but there was nothing. When they got back home, they leaned the trailer against the house and the following morning Betty's dad noticed a strip of skin, some blood and a bunch of hair snagged on the corner of the trailer.

Betty's late mother, who never threw anything away, kept the material in an envelope, which Betty took out of her handbag and placed on the table between us. She told me that about ten years ago her mother had sent about half of the sample to Montana State University but the labs there were unable to recover any DNA. With great care I removed about half of the hairs that remained in an envelope, including a small piece of attached skin, and slipped them into an evidence bag.

Each hair was about 18cm long, red-brown and fine. Some were largely straight, while others were faintly wavy. By the time I met Betty, I had spent enough time with the hair experts at the Fish and Wildlife lab in Oregon to know that this regular wave was a common feature of bear underfur. I thought that bear would be the likely species identification, but I was wrong. The lab identified sample #25212 as another canid, either a wolf, coyote or domestic dog. I asked the lab to double-check as I was surprised at the result, but Terry told me that dogs too sometimes have a wavy underfur. With mixed feelings I told Betty about the results of the DNA analysis. She took it well and was relieved that after more than forty years the mystery was now solved. She didn't see how her parents could have mistaken a dog for a bipedal sasquatch, but she suggested that perhaps what they had seen was a man and what they had hit was his dog. Was this another case of man in a monkey suit? Perhaps so, I thought, and left it at that.

One result I did have before meeting the donor again was from Marcel, the Lummi Indian. I had news of the identity of the sasquatch hair attached to the stick the creature had thrown into Marcel's back yard. Sample #25113 was a single hair, light red and straight. The lab had identified that sample too as having come from a canid. Marcel's reaction to the news was very relaxed. He then told me that next door lived a golden retriever. He had always thought of it as 'an indoor dog' but had recently found it running around in his back garden. I winced when I thought of the technical refinement of the DNA extraction, the enormously expensive equipment and the sophisticated bioinformatics required to obtain the sequence, not to mention the billions spent on the GenBank database, that had in this instance all been harnessed to identify the hair of the dog next door.

Sample #25093 came from Derek Randles, the landscape gardener and sasquatch-hunter from the Olympic Peninsula that we met in an earlier chapter. He donated two samples, one from Harstine Island in Puget Sound on the landward side of the Olympic Peninsula, which also turned out to be from a canid, while his other sample (#25086) collected near Gray's Harbor, Washington on the Olympic coast was from a cow.

Other than these domestic animals or their feral analogues, the lab list also showed me that several hairs had come from *Ursus americanus*, the black bear. These included, rather to my surprise, the bunch of hairs in sample #25104 collected by Vietnam veteran Dan Shirley and his buddy Garland Fields from the Sierra Nevada. They had been collected in winter, when bears were supposed to be hibernating. However, I was later told that black bears never really hibernate, they just doze and do venture forth from time to time. Certainly when I examined the hairs at the Fish and Wildlife Forensic lab at Ashland they didn't look like bear hairs to me. They were very pale, though I knew black bears could vary enormously in the colour of their fur. Neither did they have the characteristic wave that I associated with bear underhairs – though mistakenly so in the case of Betty Croft's trailer sample. Nonetheless, the DNA result was unambiguous. Sample #25104 was from a black bear. When I asked Bonnie Yates, Ashland's most experienced hair specialist, to have a look at Dan's sample she focused the microscope up and down, moved the stage so she could examine the hair from the root to the tip, thought about it for a few minutes then gave her opinion. It was, she thought, probably from a bear.

Another sample for which I had high hopes came from Dr Mike Amaranthus. He had left it with Peter Byrne for me to collect when I visited him in Oregon. Dr Amaranthus worked as a field scientist at the Oregon National Primate Research Center at Beaverton for many years. In 1984 he was working in the woods on the Little Chetco River, about thirty miles from Brookings on the Pacific coast of southern Oregon, when he came across a round depression of interwoven branches at the base of a giant Douglas fir. Obviously some kind of nest, he thought. Giant footprints surrounded the tree and a lot of hairs were caught on the trunk of the fir, which he collected. He and his buddy hung up engineering flags to mark the spot. When they returned the next day, the flags had been torn to shreds. At first he thought it might have been a bear, but concluded that a bear would not have had the dexterity to interweave the branches so intricately to make the nest. He sent all the hairs but one to the University of California in Berkeley but they couldn't identify them. From a microscopic examination the main suspects, bear, wolverine, coyote, even human, were excluded.

The only remaining hair had been mounted on a microscope slide under a thin glass coverslip. When Peter gave it to me, and I saw it was on a slide, I hesitated. I was reluctant to risk damaging or destroying the last remaining hair in the process of releasing it from the cement that held it under the coverslip. Fortunately my colleagues in the pathology department back in Oxford knew exactly what to do. They gave me some xylene, organic solvent, and told me to soak the slide in it overnight. The next morning the cement had dissolved, the coverslip had fallen off and the hair, about two inches long, of medium thickness and a deep red-brown, was floating free in the xylene. I already knew that xylene would not interfere with the DNA extraction as it is one of the treatments the lab uses to clean up the hair shafts. I sent the hair off straight to the lab that day and within the week the report was back. Sample #25202 was from *Ursus americanus*, a black bear after all, but one with remarkable construction abilities judging by Dr Amaranthus's description of the nest.

The other Bigfoot result that I was very keen to see was from Justin Smeja's 'Steak', which was our sample #25106. 'The Steak', you may recall, was a piece of tissue that Justin dug up at the site where he had earlier buried the young sasquatch that he shot and killed. I would class this as a second-order sample only, in that he had seen the creature at that spot, but could not be sure it was from the creature he killed. It was not a first-order sample, which the blood on his boot would have been, had we been able to identify it. 'The Steak' did not look like a bear sample to me. The hair was short, though I later discovered it may have been cut, and very light in colour. Nonetheless Ashland's other hair expert, Cookie Smith, thought this probably was from a bear, but couldn't say with much confidence. But that was what the lab result showed as well. Sample #25106 was another *Ursus americanus*. Another black bear.

A portion of 'The Steak' had previously been sent to Dr Melba Ketchum's Sasquatch Genome Project with, as noted earlier, mixed reports being fed back to Justin Smeja. Nonetheless, she does identify the mitochondrial DNA from 'The Steak' in her on-line publication as human. The issue circulating in the Bigfoot world was whether the human DNA she found on 'The Steak' had come from the sample, in which case Justin may have shot a person, or was it just contamination once again. Bear DNA had been found on 'The Steak' by another lab, but they had not used such a careful decontamination protocol as ourselves. It left the question open whether the bear DNA might have come from saliva left by a black bear as it scavenged the sasquatch carcass. Now the result was clear. 'The Steak' was from a bear and the human DNA recovered by Dr Ketchum was most probably the result of contamination.

Next, I had the uneasy task of conveying the lab result to Justin in front of the cameras. I had reminded him that he was under no obligation to do this, but he agreed all the same. The location was a jolly bar, the Red Frog, near Colfax, California and I broke the news as gently as I could. His reaction was enormous disappointment, which was only to be expected, but not primarily because he would no longer be credited with obtaining the first genuine proof of sasquatch. What really upset Justin was that he thought everyone would now assume he had lied. That was an overreaction. Given that the animal he shot had been buried for over a month and had probably been found and eaten by forest scavengers long before he returned in an attempt to locate the body, the DNA result had not shown he had lied at all. It only showed that 'The Steak' was from a bear. I pointed this out to Justin on camera, but it was cut.

Reaching now the bottom of the list of lab results, it was clear that the North American hair samples had come from very ordinary animals living within their customary geographic range. Out of the eighteen hair samples attributed to Bigfoot, five had come from black bears, four from canids, either wolf, coyote or domestic dog, three from cows, and one each from horse, deer, raccoon, porcupine, sheep and human.

All the other hair samples contributed to the project had come from Asia, mostly from the Himalayas or Russia, but with one exception. At the 'Weird Weekend' in Devon before I started the project, I had listened to a talk by Adam Davies, who I later got to know well. It was Adam who joined Lori Simmons and me on our nerve-racking mission to tempt the Big Guy out from under his tree. In his day job Adam is a local government officer in Manchester, where he works for the UK Border Agency. I was amazed to hear that he and a handful of friends use their holidays and their savings to go off on adventures to find 'animals unknown to science'. That sounded terrific. I could picture Adam returning to work after one of these trips and comparing notes with his workmates.

'Where did you go for your holidays?' I can hear him enquiring.

'Oh, you know, we went to Devon again. We always stay in the same place. It's comfortable and we can take the kids to the beach, when it's not raining that is. How about you?'

'Actually, I've just come back from Sumatra. I was there in the jungle looking for the *orang-pendek*. It was very hot and humid, and one of my mates went down with dysentery, but overall it was great.'

The *orang-pendek* was the subject of Adam's talk at 'Weird Weekend' and it was listening to him that made me realise that he and other cryptozoologists could really do with some help with their DNA research. Adam had sent *orang-pendek* samples off to a lab in the US but he never received any proper reports. On the face of it, the *orang-pendek* is the most plausible of the anomalous primates. It is quite small, standing about three feet tall, comfortable walking on two legs and covered in long, grey fur. As with other anomalous primates, there is a wealth of folk-lore surrounding the *orang-pendek*. Like the Nepalese yeti or the *migoi* of Bhutan, the indigenous Sumatrans take its existence as a given. What makes the *orang-pendek* attractive as an anomalous primate is its plausible link with a small hominid that lived in the island of Flores, not far away to the east.

In 2003 an excavation in Liang Bua cave on Flores by the Australian archaeologist Mike Morwood found the remains of a small hominid, which instantly attracted the nickname of 'the Hobbit'. There was the usual agonising, reminiscent of the first Neanderthal discovery in 1856, as to whether these fossil remains were those of a new human species or merely a stunted *Homo sapiens*. However, the present consensus among palaeontologists is that 'the Hobbit' is a genuinely new species of hominid and it has been given the scientific name *Homo floresiensis*. The fossils were carbon-dated to 19,000 years BP, so they would have been contemporary with modern humans and may have encountered them *en route* to Australia. Unfortunately, the remains are not well preserved, having been poaching gently in the tropical heat for thousands of years, and no DNA has yet been recovered. The two unsuccessful attempts were several years ago and I would have thought it was time for another try using the lessons learned from the Neanderthal and Denisovan work we encountered earlier.

Could the *orang-pendek* of Sumatra be from a surviving population of *Homo floresiensis*? An eminently plausible hypothesis, and reminiscent of Porchnev and Heuvelmans' theories about the origins of the Caucasian *almasty*. Unfortunately *orang-pendek* samples are few and far between, and as we have seen, crypto-zoologists have not fully appreciated the potential benefits of genetic analysis of hair. Nonetheless I did get a few hairs from Adam attributed, though not confidently, to *orang-pendek*. They were found caught in the bark of a tree about three feet from the ground close to a possible *orang-pendek* footprint. The right height, but unfortunately the wrong creature. The lab report identified sample #25044 as coming from *Tapirus indicus*, the Malaysian tapir. Not an *orang-pendek* this time, but another success for the lab, which had no idea the hair was from Sumatra. I now felt even more confident that we were getting fantastically accurate results.

There were nine Russian *almasty* samples. Four had come from Michael Trachtengerts at the Darwin Museum, two were from Igor Burtsev and three had come via the *Sun* newspaper. After giving my seminar in the Darwin Museum I sat down with Michael, Igor and the third member of the trio, Dmitri Bayanov, to explain the results. We started with sample #25042 from Michael Trachtengerts that I had suspected might be glass fibre of some sort and had those suspicions confirmed by Ed Espinoza at Ashland. Michael looked at this notes and identified the sample as the one from the military barracks in Archangel where two *almasty* had taken shelter from the cold in the roof space before coming down and sitting with the soldiers. The sample itself came from the chair the adult had been sitting on before they both left. There was now an explanation for the presence of glass fibre. It had come from the insulation in the loft.

The other Russian samples were all *bona fide* animal hairs and the lab had identified all of them. Sample #25039 was a single hair, quite thick, dark brown and 9cm long. It had been collected from the Kabardino-Balkarien district of northern Caucasus by Jeanne-Marie Koffman in the 1970s. The hair was found clinging to a thorn bush on the Malka Valley where there had been several encounters and numerous footprints. Having heard what a remarkable woman Jeanne-Marie Koffman was, I was very pleased to have had a sample from one of her expeditions, and also that the lab had managed to get a clear result from a single hair that was at least forty years old. Less gratifying for those around the table was the news from the laboratory. Sample #25039 was from *Equus caballus*. Another horse.

Another of Michael's samples (#25040) was also from a Koffman expedition to the North Caucasus, this time to Dolina Narzanov, the valley of the Narzan River. It was found at a campsite used by the expedition in August 1976, close to a track line of twenty-three *almasty* footprints, which were photographed and cast in plaster of Paris. There were three hairs, each about 3cm long, of medium thickness and a red-brown colour. Once more the lab result did not match the high expectations of the Moscow trio. Sample #25040 was from *Bos taurus*, a cow.

Was there going to be better news for the hard-working and persistent Russians in the third and last of Michael Trachtengerts' samples? Sample #25041 was also from the North Caucasus, but collected in 1994 near to the town of Tyrnaus in Kabardino-Balkarien. The hair was found snagged in tree bark at a place often visited by a family of *almasty* well-known to locals. There was disappointment again as the lab report identified #25041 as coming from *Equus caballus*. Another horse.

Igor Burtsev had sent me another hair sample (#25027). This one came from Siberia rather than the Caucasus. Igor had heard about a sighting in the Tashtagol region of Kemorovo province five hundred kilometres southeast of Novosibirsk. The four hairs, each about 4cm long, of medium thickness and dark brown in colour, were collected in November 2011 by the Kungushev family close to a set of large footprints in the snow. The Kungushevs had followed the prints across a road, through a meadow and down to a small creek. Here the creature appears to have leaped across before climbing up a steep wooded slope on the other side. The hairs were caught in branches at the bottom of the slope about five feet from the ground. When he heard this report, Igor Burtsev was sufficiently impressed by its authenticity to fly to Tashtagol and inspect the site himself. On both sides of the track going up the slope he found broken twigs and branches as if a large animal had passed through. It certainly had. The lab identified #25027 as coming from *Ursus arctos*, a brown bear.

Unlike Michael Trachtengerts' dignified acceptance of the results of lab tests on the hairs he had donated to the project, Igor Burtsev was stridently insistent that the Tashtagol result could not be right. He did not dispute the result of the DNA analysis, but he was so sure that the hair was from an *almasty* that he thought he must have mixed up the samples and given me the wrong one. He went to his office and came back with a small glass jar containing two hairs, also dark-brown, and said he was absolutely sure these were the real thing. Mixing up samples is easier than you might think, so I did send one of the hairs off for analysis but it came back with the same result. *Ursus arctos*.

It is at times like this that I reflect whether testing these hairs was really such a good idea. The three men had worked for over forty years on the *almasty*. True, all three were very interested in collecting folk tales, especially Dmitri Bayanov, and they had spent a great deal of time writing down the accounts of hundreds of eyewitnesses. True also was the fact that, when these hairs were collected, their importance as an infallible means of identification was not appreciated. Of course, the three men hoped passionately that the hairs would vindicate their years of dedication. More than that, they really *expected* the results to prove at last that the *almasty* was a surviving Neanderthal or, at the very least, an unidentified primate. I would have liked that too. But the fact is that the hairs actually came from a mundane selection of very ordinary animals. The three men were understandably quiet and crestfallen after I told them the news. It struck me then that it was a good thing that the finality of the results could not be argued about. Had I given an opinion rather than a factual truth, there would have been room for doubt and debate. Any ambiguity, thank heavens, was avoided by our DNA results, but the truth can be a bitter pill to swallow. Would it have been better not to have offered the medicine in the first place?

However, I knew something which at that point they did not. I had the results of my analysis of the Zana case. And, as I will tell you very soon, that made everyone around the table forget the disappointment over the hair results.

But before we leave Russia, there is one more story to tell, and more results to report. Early in 2013 I got a call from Emma Little, the health and science editor of the *Sun* newspaper. One of the paper's news editors had been approached by a Russian offering a genuine yeti hair at a cost of several thousand pounds. The editor approved the payment and the hair was sent to Edinburgh Zoo for DNA analysis. This destination was chosen because, as you may recall from Chapter 19, one of the zoo's scientists, Dr Rob Ogden, had recently analysed the infamous Pangboche Finger. I had already arranged to visit Dr Ogden to ask him about the finger, so I took the opportunity to enquire about the yeti hair from the *Sun*. It had never arrived, and appeared to have been lost in transit. Emma Little's editor had not given up having acquired, also at considerable expense, three more yeti hairs from the mysterious Russian source. Emma asked if I would test these, I agreed and the lab got results from all three. They turned out to be from *Equus caballus*, another horse, *Ursus americanus*, an American black bear and, last of all, *Procyon lotor*, a raccoon. The oddest things about this altogether odd episode was that the last two, the black bear and the raccoon are not Russian at all, but natives of America.

Emma Little wrote the article under the headline that some might think fitting for all my hair sample work so far. It read: 'Prof Spoilsport disproves Yeti.'

Postscript

I received news from the laboratory of one final sample only three days before the final manuscript for *The Nature of the Beast* was due at the publishers. It is an intriguing result that I want to tell you about. The hair was donated by Dr Henner Fahrenbach, who more than anyone else has made a speciality of the study of sasquatch hair and whose identification guidelines we covered in Chapter 9. Henner Fahrenbach has been fascinated by microscopy all his life – but let him tell his own story to you just as he did to me:

'I was for my entire career in charge of the electron microscopy laboratory at the Oregon National Primate Research Center. I lived and breathed microscopy from childhood and eventually got paid for it. I started as a marine invertebrate zoologist, shifted during graduate school to invertebrate histology, and subsequently by way of postdoctoral work at the Harvard Anatomy Department to electron microscopy. I worked broadly on different aspects of anatomy and histology, from low level crustaceans through chelicerates (*a large group of invertebrates which includes spiders, scorpions and the extinct trilobites*) and insects to lemurs and man, both normal and pathological aspects of their organ systems, though my specialty remained the visual system, particularly that of the horseshoe crab (*Limulus polyphemus*).'

He explained how he became interested in sasquatch, aside from living in Oregon where it is hard to avoid being caught up in the local enthusiasm for the creature.

'I was always a wilderness backpacker and mountain climber

and had a healthy, though private interest, in the sasquatch from the 1960s on (pre-movie!) (*Here he is referring to the 1967 Patterson-Gimlin film.*) But I only started engaging in active field research in the 80s. I have personally found footprints, heard their cries and screams numerous times, got close enough to smell one, and interrogated hundreds of first-hand eyewitnesses.'

His final conversion came when he carried out a statistical analysis of Bigfoot sightings all over America which was published in 1997.[1] He retired from teaching in 2007 and moved from Oregon to Gold Canyon, Arizona.

I had been trying to contact Dr Fahrenbach for some time as I had read in Dr Meldrum's book, *Sasquatch. Legend meets Science*[2] that Dr Fahrenbach had some hairs recovered from the scene of the Skookum cast where, as we saw in Chapter 14, a sasquatch had been resting and had left an impression in the ground. The hair had been sent for analysis to Dr Craig Newton at the University of British Columbia but, as we also heard, he was unable to recover any DNA other than his own. I thought it was worth another crack at the Skookum hair using the rigorous decontamination steps in our analytical protocol that had managed successfully to circumvent the 'curse of contamination'.

Dr Fahrenbach had been unwell, so I did not hear from him for a long time, but when he did reply he thought it a much better idea to test another hair altogether, one that satisfied his sasquatch criteria and that he had identified from its microscopic appearance as having come from a primate. This is the famous Walla Walla sample which had been collected by Paul Freeman in 1997. The more Dr Fahrenbach told me about the provenance, the more determined I became to analyse it.

'Alerted by some bow hunters, who said they had seen "an awful lot of tracks and torn-up stumps" on a forest road twenty five miles from Walla Walla, Washington (*which lies just north of the Oregon border*), Paul Freeman alerted two friends. These were Wes Sumerlin, an elderly Yakima Indian and an experienced tracker, and Bill Laughery, a retired game warden and forester. The arrived at the site at 9 a.m. the next morning,

4 August 1997, and proceeded to track three sets of footprints measuring sixteen, twelve and eleven inches respectively. The last track was missing the large and small toes of the right foot.

'After nine hundred feet of tracking the footprints (all distances were subsequently verified by tape measurements or paced off) they encountered a suspicious smell. Sixty feet further on they came upon several balsam firs that were broken four feet above the ground, twisted through 180 degrees, and measured between two and a quarter and three and three-quarters of an inch at the break. The breaks were very fresh since the cambium (*part of the exposed internal surface of the branch*) was still visibly oozing sap. Where the trees had been gripped, about eight inches to either side of the break, the bark looked darker by having been compressed.

'Two of the trees had substantial amounts of hair caught in the splintered wood and these were collected. Near the broken trees a number of wood violets had been pulled out of the ground, their stems eaten and the few top leaves discarded in small heaps.

'At this point the three trackers separated slightly, Paul Freeman walking along a lumber road, while Bill Laughery moved towards the Mill Creek boundary with Wes Sumerlin trailing him by twelve to fifteen yards. About forty yards from the twist-offs (*i.e. the broken trees*), Wes saw a flicker of motion in the deep timber with dappled sunlight breaking through the foliage. Then Bill Laughery walked twenty yards further ahead and saw something standing still in the dense forest. He alerted Wes, who approached him. Bill then examined the creature from a distance of just under ninety feet through 7 x 26 binoculars for about five minutes, the animal standing stock still throughout.

'It was a light tawny brown in colour, facing slightly away towards where Wes had been. Bill could observe its brow ridge and a sagittal crest of about one and a half inches that ran across the top of its head. The remainder of the face was largely obscured by the contrasting light and shade, and by its facing partly away from where Bill was standing. Its chest showed no mammary glands and no genitalia were visible. Its coat was longer than that of the female in the Patterson-Gimlin film, appearing to be longest on the upper arm and back, shorter on the trunk. The attention of the sasquatch remained directed elsewhere and it did not look at the observers. When Bill momentarily turned to Wes to whisper something, Wes saw the creature move away without a sound.

'Bill and Wes followed the track for three hundred feet. At around the same time, another animal leaped, in one bound, fifteen feet across a jeep track in front of Paul Freeman. All three trackers could hear animals moving through the trees and one of them gave a sharp whistle from a nearby canyon. At this point the trackers sat down and waited in silence for about fifteen minutes. Then the smallest of the three sasquatch broke cover and headed for a fence about a thousand feet away, leaving behind a set of twelve-inch footprints.'

As sasquatch encounters go, they don't get much better than this. There were three witnesses, they saw at least two different sasquatch, one through binoculars whose 26mm lenses would have given an excellent image at the range used, even in the gloom of the forest. Moreover, they actually saw the animal that left the twelve inch footprints as it made a dash for the fence. The hair was found caught on the branches of the balsam fir that had been so powerfully twisted off. So we can be sure that the hair belonged to the creature that had broken off the branches. We cannot say for sure that these were the same creatures that the trackers saw soon afterwards (remember the twist-offs were fresh) but there is obviously a good chance that they were. Even if they were not, to identify the creatures who twisted off the balsam firs would be a great step forward.

In our correspondence about the Walla Walla sample, Dr Fahrenbach told me how his precious collection of hairs had been gradually eroded, having been sent to numerous laboratories but never returned with any concrete results. So I was privileged to be offered the last of the Walla Walla hairs. It was 10cm long, but Terry Melton's lab managed to complete the DNA analysis on a section of hair shaft only 2.5cm long. I know I keep repeating myself when I stress how superbly the lab succeeded in coaxing DNA sequences from such small samples, but the ability to do so consistently and accurately has been the key difference between this project and all the others that have gone before. The 12S RNA sequence from the Walla Walla hair was quite clear. Dr Fahrenbach had been right in identifying its primate origin. This hair was human.

I found this result very exciting indeed and immediately asked the lab to do the next test on the extracted DNA that would tell me whether it was from a *Homo sapiens* or another human species. Remember that the 12S RNA sequence is only a rough guide to species identification. We have seen that among canids, for example, it does not differentiate between a wolf, a coyote or a domestic dog, and I have explained the good reasons why this is so. The same goes for humans. From the 12S RNA result alone the Walla Walla hair could have been from a Neanderthal, a Denisovan or another as yet unknown species in the genus *Homo*.

The test I ordered was on the highly variable control region we met in Chapter 13. The sequence of this region can easily distinguish between a *Homo sapiens* and a Neanderthal, and any other collateral hominid that had left its hair behind, snagged in the broken balsam firs. The only other Bigfoot sample that required this advanced treatment with the control region was sample #25072 found on the window ledge of a cabin in Oklahoma after the sasquatch attack. As we saw earlier on in this chapter, sample #25072 proved to be a modern human with a mitochondrial control region sequence indicating a European ancestry.

It seemed an age before the lab reported the control region sequence of the Walla Walla hair and I subsequently learned from Terry Melton that it had been particularly difficult to work with. Her lab technicians had been forced by the relatively poor condition of the recovered DNA to build up the control region sequence from a number of smaller segments. Each segment had to be analysed twice, which added to the time it took to get a result.

Terry knew my book deadline was looming and worked her staff hard to get the control region to me in time. I am glad she did because it is intriguing. The DNA sequence from the Walla Walla hair was very unusual indeed.[3] I had not seen it among my reference databases containing several hundred thousand sequences. Nonetheless it had some similarities to a modern human, and was certainly not from a Neanderthal. Eventually I found a match in a rather obscure database from Central Asia. The Walla Walla sample matched an individual from Uzbekistan!

How on earth could that be explained? I have not had long to think about it, but my immediate thought is that I find it very difficult to reconcile this result on the Walla Walla hair with the impressive provenance provided for it by Paul Freeman and his companions. The hair was caught in the splintered wood of a tree whose branch had been twisted off with tremendous force. Had the Walla Walla hair been found lying on the ground in the vicinity of a Bigfoot 'experience' then a loose human hair is always a possibility. But this hair *must* have belonged to whatever creature broke the branch, even it was not one of the sasquatch the trackers saw in the vicinity soon after. So as things stand, it is a complete mystery.

There are obvious things to do, the first of which is to repeat the analysis with another of the Walla Walla hairs. Dr Fahrenbach gave me the last of his 'free-standing' hairs, but he tells me he has others mounted on microscope slides, just as you would expect from a microscopist. I know from the Amaranthus hair (sample #25203) that I am able to remove a hair from underneath a microscope coverslip without compromising the DNA analysis, so I hope to be able to do that in the near future. But what a tragedy that so many of the Walla Walla hairs ended up in labs that Dr Fahrenbach tells me often didn't even bother to acknowledge receipt, let alone feed back the results of a good DNA analysis.

The Walla Walla hair result is the most intriguing from among my North American samples. I scarcely think I can claim to have identified the sasquatch as a feral Uzbek, but that is the closest I have managed to get at the moment.

28

The Snow Bear

Three species of bear live in the Himalayas. The shaggy-coated sloth bear, *Melursus ursinus*, inhabits the forested foothills up to five thousand feet, feeding mainly on termites and ants. They may be the smallest of the Himalayan bears, though adult males can reach three hundred pounds, but sloth bears are the most dangerous of the three native species. They have killed or injured more people than the other two because they inhabit the more populated low-altitude regions. Though not naturally aggressive, sloth bears will attack if surprised and are equipped with long curving claws on their front legs for digging out termites that can disembowel a man with a single slashing blow. Higher up the slopes, to an altitude of twelve thousand feet, is the habitat of *Ursus thibetanus*, the Asiatic black bear. This bear is larger than the sloth bear, with males weighing up to four hundred pounds, and is omnivorous. Both sloth and Asiatic black bears have a crescent-shaped white mark across the chest which distinguishes them from by far the largest of the Himalayan bears – *Ursus arctos*, the brown bear. This huge animal, with males up to eight feet tall and weighing a thousand pounds, lives at even greater altitude than the other two species and has been seen at over fifteen thousand feet, well above the tree line. Although all three species of bear can walk for short distances on two legs, it is the brown bear's sheer size and its high-altitude habitat that makes it the most plausible of the native bears to be confused with the yeti and *migoi*.

There is a lot of uncertainty surrounding the precise taxonomic definition of the brown bear as a true species. In the past, brown bears have been separated into many geographical sub-species, some like the Tibetan bear and the Blue bear being natives of the Himalayas. However, the most recent consensus among taxonomists is that brown bears the world over are members of a single species in the sense, as we covered in Chapter 11, that we now know that they can, and do, successfully interbreed in the wild. This modern view groups brown bears across the complete size range from the comparatively small European and Syrian bears to the giant Kodiak grizzly of Alaska under a single umbrella species: *Ursus arctos*.

When I received the result from the yeti mummy discovered by Christophe Hagenmuller in Ladakh, I was not too surprised that the mitochondrial 12S RNA sequence identified it as a bear. Sure enough, when I ran the sequence against the GenBank database most of the matches were with specimens of *Ursus arctos*, the brown bear. However, the closest match in GenBank came from another species of bear altogether, and it was not one of the other natives of the Himalayas. The sequence in GenBank that matched the Ladakh yeti sample came instead from *Ursus maritimus*, the polar bear.

I was utterly amazed at this result. How could a polar bear be living so far from the Arctic? Although the Ladakh yeti mummy was not white, but a golden brown, I knew that white bears had occasionally been seen in the Himalayas. In his book *The Valley of Flowers*, the explorer Frank Smythe relates how, 'The Sikh surveyor who I had met in the valley was reported by the Postmaster of Joshimash as having seen a huge white bear in the neighbourhood of the Bhyundar valley (*in the western Himalaya*)'.[1] The zoologist Bernd Brunner mentions two occurrences of white bears in East Asia in his monograph *Bears: A Brief History*.[2] One is from the record of a royal shipment from China to Japan of two 'white bears' and seventy 'white bearskins' in AD 685. The other is from the journal of Marco Polo, the Venetian explorer, who saw one on his travels through Mongolia in the early thirteenth century. Brunner suggests these reports might refer to polar bears, or alternatively to giant pandas, which were unknown in the West at the time.

When I looked into the details of the genetic match I was in for another surprise. Each GenBank accession is linked to details of the specimen from which the DNA sequence was recovered, and to any associated scientific publications. The paper linked to the polar bear sequence matching the Ladakh mummy had been published in 2010 in the prestigious US journal *Proceedings of the National Academy of Science*.[3] It carried the intriguing title 'Complete mitochondrial genome of a Pleistocene jawbone unveils the origin of polar bear'. Its principal author, and head of the collaboration of American and Norwegian scientists that carried out the research, was Charlotte Lindqvist. The title itself hinted at the prospect that the Ladakh mummy DNA matched an ancient polar bear rather than a modern specimen.

I quickly got hold of a copy of the Lindqvist paper. Owing to its iconic status as a symbol of climate change, the polar bear has received a lot of attention from biologists in recent years. Indeed the paper opens with the sentence: 'The polar bear has become the flagship species in the climate-change discussion'. It goes on: 'However, little is known about how past climate impacted its evolution and persistence, given an extremely poor fossil record', before explaining that the research group had managed to obtain a DNA sequence from one of the very rare fossils, which was the focus of the paper. The scarcity of polar bear fossils is easily explained because the animals mostly live and die on the sea ice and even if the remains are not consumed by scavengers, including other polar bears, there is little prospect of them ever being preserved. For fossilisation to occur a polar bear must die on land and its remains be rapidly covered by sediment that subsequently freezes.

The rare fossil that the Lindqvist team had analysed was a jawbone, and had indeed been found on land. It was then stored in the vaults of the University Museum of Natural History in Oslo, Norway where a member of Lindqvist's team discovered it. The jawbone had been excavated at Poolepynten on Prins Karls Forland, a narrow strip of land on the far western edge of the Svalbard Archipelago in the Arctic Ocean, five hundred miles north of the Norwegian mainland. This collection of remote icebound islands is still home to a large concentration of polar bears and is usually reached by air from the city of Tromsø, high up in northern Norway, well above the Arctic Circle. I visited Tromsø a few years ago and remember being shaken by a stark notice on the airport gate leading to the Svalbard plane reminding passengers that polar bears are not the cuddly creatures most imagine, but dangerous carnivores. The written warning was accompanied by a vivid photograph of two polar bears, their heads covered in blood, feasting on a carcass. A tourist possibly, but more likely a seal.

The Svalbard jaw from the Oslo museum was well preserved, as you might expect from a fossil that had been permanently frozen since shortly after its death, and there was little to indicate how old it was. The first step in the investigation was to carbon-date the specimen. This is done by measuring the amount of the slightly radioactive isotope carbon-14 that remains in a specimen. It decays slowly over time once the animal has died, so that as time passes less and less of the isotope remains in the fossil. The less carbon-14 there is, measured relative to the stable isotope carbon-12, the older is the fossil. This works well up to about 45,000 years, after which there is too little carbon-14 left to make an accurate age estimate.

To everyone's surprise, the Svalbard jaw was too old for carbon-dating, meaning it must have been at least 45,000 years old. However, there were sediments surrounding the fossil that could be dated by a technique called infrared luminescence. This technique works by heating a sediment sample, whereupon the radiation built up in mineral crystals since they were last exposed to light is released and measured. The more comes out, the older the sample. These are not direct measurements on the fossil, and of course the sediments may have been washed in from elsewhere. However in the case of the Svalbard jaw, infrared luminescence gave a much older estimate than the carbon-dating limit of 45,000 years, estimating the age of the jaw at 110,000 to 130,000 years. That was an exciting discovery for the biologists because it pushed the date at which polar bears were believed to have emerged as a new species further back into the past.

So what to make of the result from the Ladakh yeti mummy? How could the DNA from a modern creature from the Himalayas match that of an ancient polar bear? The first, rather prosaic explanation is that the Ladakh specimen is a regular brown bear with a chance mutation in its mitochondrial 12S RNA gene which made it match the DNA from the ancient polar bear. Certainly it would only take a single mutation from a modern brown bear to do that. At the DNA base five in from the end of the fragment that we sequenced, a 'C' in the brown bear is replaced by a 'T' in the ancient polar bear and in the yeti mummy. It is possible that this mutation from a 'C' to a 'T' happened independently in the Ladakh specimen, or more probably in one of its matrilineal ancestors. Given the glacial mutation rate of the 12S RNA gene, this is not very likely. It would also be unlikely for an independent mutation in the Ladakh specimen's ancestry to have occurred at the one place that created the match to the ancient polar bear and not somewhere else along the mitochondrial fragment that we sequenced. The only way to find out is to get more mitochondrial sequence from the yeti mummy and see how it compares to *Ursus arctos* and the ancient *Ursus maritimus*. As I write, this work has started, but the hair is old and not in great condition, so it is proving to be a difficult technical challenge. Nonetheless we should eventually be able to sequence the entire 16,800 bases of the Ladakh mummy's mitochondrial

genome and explore its genetic relationship to the ancient polar
bear much more thoroughly.

Let us suppose for the moment that the DNA connection with
the ancient polar bear is real. How can it be explained? It would
follow that the Svalbard and Ladakh bears are closely related. It
is unlikely that the Ladakh yeti was a direct matrilineal descendant
of the Svalbard polar bear. Far more probable is that the two
specimens both inherited their mitochondrial DNA from a
common ancestor that must have lived at least 40,000 years ago.
In any event there would be a direct matrilineal link between
the Ladakh yeti and the Svalbard polar bear.

Does the link between the Svalbard and Himalayan specimens
mean that the ancestors of the Ladakh yeti/bear had travelled all
the way from the Arctic? That's certainly a formal possibility. The
journey need not have been a recent one and may even have
been undertaken when the climate was quite different. Even so,
the ancestors of the Ladakh yeti/bear would have needed to switch
their diet from maritime seals to something altogether more
terrestrial. These ursine wanderers would also have changed
colour from the white of the modern polar bear to the golden-
brown of the Ladakh specimen. None of this is impossible. Under
the circumstances you would expect evolutionary adaptations to
life on land, of which a colour change is one.

There is another explanation. It is well known that in captivity
brown bears and polar bears can interbreed to produce fertile
offspring. In Chapter 11, I emphasised how rare true hybridisation is in the wild, but it now appears that polar x brown bear
hybrids may be an exception. This intriguing possibility comes
from research done on the bear population of the Admiralty,
Baranof and Chichagof (ABC) islands of southeast Alaska's
Alexander Archipelago.[4]

Here the bears look and behave like brown bears, but they
have the mitochondrial DNA of polar bears. Not the ancient
type like Svalbard, but that of modern polar bears. This implies
that there has been interbreeding between the two bear species
at some time on the past, with at least the female participant

being a polar bear. When the nuclear DNA from the ABC bears was sequenced and compared to browns and polars, it revealed that most of the genome was closely matched to brown bears with only a minority coming from the polar bear ancestral female. This shows that the ABC bears are true hybrids, but also raises once again the semantic definition of a species. If in the case of the ABC bears there is successful interbreeding in the wild, then this must mean that formally they are both in the same species. Semantics perhaps, but manna from heaven for taxonomists. They will be able to argue about it for years.

The Ladakh yeti/bear could also be descended from ancestors who, rather like the ABC bears of today, were hybrids between Svalbard polar and mainland brown bears. Interestingly, there are indications that as polar bears lose their hunting grounds on the ice they are moving inland into territories normally occupied by brown bears. On the rare occasions when a meeting between polar bears and brown bears has been observed, there is usually a fight which, being larger, the polar bears usually win. But the evidence from the ABC islands is that this aggression can sometimes turn into sexual encounters, forced or otherwise, with the birth of hybrid cubs being the result. A good quality nuclear DNA sequence should reveal, if the Ladakh specimen is a hybrid, the relative proportions of brown and polar in its genome. This work is also under way, though while the only source is the forty-year-old hair shafts, it is proving to be even more difficult than obtaining a full mitochondrial sequence.

The Ladakh result was completely unexpected, and raises a number of questions, and not just about how polar bear DNA got to Ladakh. The yeti mummy was shot over forty years ago, and there have been many yeti sightings since then. Could it be that if this 'new kind of bear' was the biological template for the yeti, there were others? That is when I remembered the Bhutan *migoi* samples that my lab had analysed ten years before. As you may recall from earlier chapters, my research student Helen Chandler had identified two of the three samples as regular bears, one as a brown bear *U.arctos,* the other as an Asiatic black bear *U.thibetanus.* The third sample could not be identified.

It matched nothing in GenBank and, in my opinion, was probably an artefact of the amplification reaction. When there isn't enough good quality DNA, the reaction often stitches random short segments together and gives a nonsensical sequence. Even though this was a long time ago, there were already enough bear sequences in GenBank to make a comparison, and the Bhutan hair was nothing like any of them. As I was digesting the surprising results from the Ladakh hair, I had an idea and contacted Helen to ask her if she could remember whether she had used the whole of the third Bhutanese *migoi* hair in the testing. We only ever had a single hair, and I remembered it had a plump root. She replied that her lab notebook recorded that the sample was used up in the extraction process. Ah well, I thought, and forgot all about it.

Well, almost all. I did have a look in the lab freezer, but found nothing that could have been the third Bhutanese sample, not surprising given that the contents have been moved several times in the interim. Some weeks later I was looking through my DNA samples for something else and noticed a box of storage tubes labelled 'N.E. Pakistan'. Inside the box were DNA samples collected in the 1990s by a medical student attached to the lab who had travelled to Gilgit in northern Pakistan. The idea behind her undergraduate project was to see if there was any genetic evidence to support the popular legend that the Burusho people of the region trace their ancestry to the soldiers who accompanied Alexander the Great during his military expeditions in the fourth century BC. There was, but that is another story.

Among the Gilgit samples lay four slightly larger tubes with yellow labels. They read 'Bhutan 1', 'Bhutan 2', 'Bhutan 3' and 'Bhutan 4'. I thawed them out and in 'Bhutan 3' I found a very short segment from the tapering tip of a hair shaft. It was less than an inch long. I must have snipped it of before I gave it to Helen to extract. I do things like that. I confirmed from Helen's thesis that 'Bhutan 3' had been the mystery sample with no sensible result and straightaway sent it to Terry's lab. I was hoping for, though not expecting, another match to the Svalbard polar bear. You can imagine my excitement and my satisfaction when I received the lab report. *Ursus maritimus*. Another match to the ancient polar bear. Brilliant!

This had to mean that the Ladakh yeti/bear was not alone. Whether you call it a yeti, a *migoi*, a polar bear or a hybrid, the creature was living in Bhutan ten years ago when the 'Bhutan 3' sample was taken from the tree nest in a high-altitude bamboo forest. Two bears. One from Ladakh in the western Himalayas and one from Bhutan in the east. Surely there must be some others in between.

It began to look more and more as if the descendants of the ancient Svalbard polar bear really had travelled to the Himalayas at some point in the past. The whole story was beginning to show an uncanny resemblance to an episode in Philip Pullman's novel *The Amber Spyglass*, in which Iorek Byrnison, king of the armoured Panserbjørn, is forced to lead his polar bears to the Himalayas because the climate in his home country of Svalbard has become inhospitable. Whether the ancestors of the yeti/bear were led there by Iorek Byrnison or came under their own steam, the Ladakh and Bhutanese yetis could be polar bears a long way from home.[5]

I returned in 2013 to South Tyrol to meet up with Reinhold Messner. It was a much more relaxing experience than our first encounter. I was greeted with a broad smile and he called me 'Bryan'. I was there to give him the news about his two yetis that I had sampled the previous year. Despite trying several times, I had not been able to get any DNA from either of them. I had examined the hairs thoroughly in the Oregon lab and they looked in bad shape, with some sort of lacquer coating sticking to the outer surface. I had picked out the four best from my collection for DNA analysis, but to no avail. The samples were quite old, having been shot in the 1930s, but others of a similar vintage had worked without any great difficulty. The only explanation I could give Messner was one I received from the chief conservator at the Muséum National d'Histoire Naturelle in Paris, whom I visited on the track of a yeti sample reputed to have been brought back from the Himalayas by the orchid collector René de Milleville in the 1990s. We could not locate the de Milleville sample, but he did give me some valuable advice about taxidermy. Apparently there are different fashions in taxidermy and the German method before the war was to bathe the entire skin in an acidic solution before tanning. Tanning prevents the skin from rotting and the acid treatment that preceded it would have destroyed any residual DNA. The method in many other countries at that time, and in Germany too these days, avoids the use of acid altogether, which I think explains my success in recovering DNA from specimens held in museums in other countries and preserved at different times.

Messner didn't take the news too badly, I thought, and really cheered up when I told him about the genetic match I had found between the Ladakh mummy and the polar bear. This really might be the 'different kind of bear' that his travels in Tibet and Nepal had convinced him lay behind the yeti legend. This could be the 'chemo'. Given that I had found the same bear at each end of the Himalayas, what did he think of the chances of finding living examples?

Another broad smile lit up his rugged face. 'I will lead the expedition *personally*,' he replied, placing great emphasis on 'personally'. And that is exactly what will happen one day soon.

29

Zana

The story of Zana fascinates everyone with even the most fleeting interest in the legends of the wildwood and the strange creatures, half-human, half-beast, that lurk in its shadows. It is by far the most absorbing account of the capture of an anomalous primate, an apeman, that there has ever been. Unlike so many other stories that depend on unsubstantiated eyewitness accounts, *hundreds* of people saw Zana the wildwoman in the forty of more years she was held in captivity. Even more unusually, there were physical remains to examine and, especially for me, genetic analyses that I could perform that might discover what sort of creature Zana really was.

We know the details of Zana's capture thanks to the work in the early 1950s of the Russian zoologist Alexander Mashdovtsev and his young associate Boris Porshnev, who went on to become the leading figure in Russian hominology. Although Zana had died sixty years previously, Mashdovtsev and Porshnev were able to find and interview several witnesses who as young children had seen Zana and remembered her well.

The location for Zana's story is Abkhazia, now a country devastated by war and uneasily positioned between Russia and Georgia on the southern slopes of the Caucasus Mountains. Densely wooded valleys radiate towards Russia in the north and to Abkhazia in the south from the spine of high peaks which climb to Mount Elbrus at 18,500 feet and stretch between the Black and Caspian Seas. Even more than most remote mountainous areas, the Caucasus has an abundance of wildman legends stretching back centuries, and it was the almost permanent base for Jeanne-Marie Koffman's expeditions in the 1960s and 1970s in search of the elusive *almasty*.

In the early 1850s, a travelling merchant visiting the Ochamchir region of Abkhazia on the southern slopes of the mountain range came across a young *almasty* by a remote stretch of the Adzyubzha River. As soon as it caught sight of him, the creature vanished into the forest. Some days later he returned with a group of hunters and their dogs. When they saw it again, the dogs were unleashed, chased it into the forest and brought it down. After a fierce struggle it was eventually subdued, tied, gagged and shackled to a log. It was clearly a female. She was held for a while in a ditch surrounded by sharpened wooden stakes, then sold on to a succession of 'owners' until she was eventually purchased by the Abkhaz nobleman Edgi Genaba and taken to his farming estate at Tkhina on the Mokva River. Here she spent the rest of her life until she died around 1890.

Mashdovtsev and Porshnev obtained detailed and consistent eyewitness descriptions of the creature, named Zana, in the 1950s.

Part-human, part-ape with dark skin (Zana means 'black' in Abkhaz) she was covered with long, reddish-brown hair which formed a mane down her back. She was large, about 6'6" tall, and extremely muscular with exaggerated, hairless buttocks and large breasts. Her face was wide with high cheekbones and a broad nose. Brilliant white teeth flashed from her wide mouth, teeth which could crush nuts and even bones. Her strength was such that she could lift a fifty-kilogram sack of flour with one hand and hold it steady for several minutes.

At first Zana was kept in a stone-walled enclosure near Genaba's house and her food was simply thrown over the wall. She dug a hole in the ground within the enclosure where she slept. Zana was a source of cruel amusement for local children, including some of the eyewitnesses later interviewed by Mashdovtsev and Porshnev. The children used to torment Zana by prodding her with sticks through a gap in the wall or pelting her with small stones. Despite these provocations, Zana became gradually less aggressive as the years passed. She was moved from her stone prison to an enclosure close to Geneba's house before finally being set free to roam the estate.

Zana hated being indoors, preferring to sleep in the open air, often lying down with the estate's water buffalo in a stagnant pool. She never wore any clothes, and even in the depth of winter spent all day and night completely naked, never attempting to cover herself against the cold. Zana enjoyed swimming in the Mokva River, even when it was in full flood, and showed her athleticism by racing Geneba's horses. She never tried to escape, and began to do menial tasks for Genaba, including grinding corn in his watermill. Though apparently not fully human, she became his slave.

One aspect of the eyewitness descriptions that I find most remarkable is that Zana never tried to speak. Though she had a repertoire of inarticulate grunts, whistles and cries, she never managed to learn a single word in Abkhaz or attempt to speak in her own language, whatever that might have been.

In many ways Zana's is a classic tale of a wild creature part-human, part-animal, captured and tamed. And so it would have remained, a story, and we would have known nothing more about Zana. What has kept the story alive is that she had at least four children with local men. The circumstances are unclear but there are tales of drunken orgies and curious men being granted access to her in exchange for money.

Naturally there has been a lot of speculation about who or

what Zana may have been ever since Mashdovtsev and Porshnev first brought her story to a wider public. Zana's case has been studied not only by Porchnev, but also by his contemporary and fellow cryptozoologist Bernard Heuvelmans, whom we have met several times already. They wrote a book together, *L'es Homme de Néanderthal est toujours vivant (Neanderthal Man is still living)*, which presents their ideas about Zana.[2] The title of the book leaves readers in no doubt as to its main conclusion. Heuvelmans and Porchnev believed Zana was a surviving Neanderthal.

Jeanne-Marie Koffman, who spent more time than anybody exploring the Caucasus both north and south of the highest mountains, summarised the many eyewitness accounts of the *almasty* she recorded during the 1960s and 1970s. They closely paralleled the descriptions of Zana obtained by Mashdovtsev and Porshnev in Abkhazia in the 1950s.

'Là où il n'ya pas de poils, la peau est noire.'

'La peau du visage est noire.'

'La peau du paume des mains est marron foncé. Sur les fesses, les poils sont absents, la peau est marron foncé.'[3]

(Where there is no hair, the skin is black . . . The skin on the face is black . . . The skin of the palms is dark brown. On the buttocks, hair is absent, the skin is dark brown.)

As I mentioned in Chapter 3, I had become interested in the Caucasus as a possible refuge for surviving Neanderthals. I had been keeping an eye out for Neanderthal mitochondrial DNA sequences in the tens of thousands of living people taking part in personal ancestry projects through my own company, Oxford Ancestors, and others. The Caucasus was one of the regions that palaeontologist Chris Stringer and I picked as the likeliest location to find modern humans who carried Neanderthal mito-chondrial DNA through ancestral interbreeding.

Two developments turned Zana's story from an intriguing folk tale, albeit substantiated by several eyewitnesses, into a case with potential for real scientific investigation. The first of these was that in 1971 Igor Burtsev located the grave of Khwit, the younger of Zana's two sons, in an overgrown graveyard in Tkhina and exhumed his body. The second was that Burtsev, and also lately Dmitri Pirkulov, have managed to trace six of Zana's living descendants.

When Khwit's skull arrived back in Moscow it was examined by two anthropologists, M.A. Kolodieva and M.M. Gerasimova. As reported in Dmitri Bayanov's 1996 book *In the Footsteps of the Russian Snowman*, Gerasimova pointed out several peculiarities in the skull, including its very large dimensions, while Kolodieva saw a mixture of modern and archaic features, the implication being that Zana had contributed the archaic elements of Khwit's skull while his father, whoever he was, was responsible for the modern ingredient.[1] When I examined and measured the skull during my visit to Moscow in the summer of 2013, it certainly was unusually large and a multivariate analysis of twenty-nine standard skull dimensions put the skull outside the range of modern human variation.

Sadly, despite several attempts, Igor Burtsev was never able to locate Zana's remains. While Khwit's skull is certainly peculiar in several respects, and is now being re-evaluated with more up-to-date techniques, it is not Zana's. Any conclusions about her origins from examining the features of her son's skull must be tempered with that important proviso in mind.

I could see a way of obtaining direct genetic information about Zana from her son's skull undiluted by his father's input. If I could recover mitochondrial DNA from Khwit's skull, the strict matrilineal inheritance would mean that its sequence was identical to Zana's. I could also see a way of discovering even more about the Zana's genetic make-up through her living descendants. Both were exciting prospects but there were several obstacles to overcome, the first of which was getting hold of something to test, which wasn't going to be easy.

Khwit's skull is now in a private collection in Moscow, having been sold by the cash-strapped Darwin Museum. Somehow Dmitri Pirkulov used his abundant charm to persuade the skull's owner to allow him to remove an incisor tooth, which he handed to me when we met in London in late 2012. He also showed me his reconstruction of the Zana pedigree and pointed out the descendants that he and Burtsev had traced. There were six of them, five descended from Khwit's brother Eshba and one from Khwit himself. Pirkulov also brought bloodstained tissues from some of the descendants, but I did not think I could do much with those and urged him to return to Abkhazia with saliva sampling kits that I could use for the tests I had in mind.

With Khwit's tooth to test, the next hurdle was the tricky matter of recovering DNA from it. The tooth, embedded in Khwit's upper jaw, had lain in the ground for twenty years between 1952, the year he died, and 1971 when Igor Burtsev exhumed his body. Though I have recovered DNA from teeth well over ten thousand years old, it is not the age so much as the conditions of preservation that influence whether any DNA survives in a skeleton, and crucially whether it is in good enough shape to generate a DNA sequence. By far the best place to find fossils with well-preserved DNA is in a cool limestone cave. Here the temperature is stable throughout the year and the alkaline conditions preserve the bone and tooth minerals, which in turn protect the DNA within. On the other hand, acid soil or cycles of heat and cold will destroy DNA very quickly. The climatic conditions in Tkhina, with hot summers and very cold winters, are certainly not ideal for DNA survival.

Though it was not easy, Terry Melton and I did manage to recover and sequence mitochondrial DNA from Khwit's tooth. We used a modification of the hair decontamination protocol to clean the surface of the tooth then powdered it in a liquid nitrogen mill and extracted the DNA from the residue. The tooth was fifty years old, but despite that it gave a yield of high quality DNA from which in a matter of days we were able to obtain a good mitochondrial control region sequence.

As soon as Khwit's sequence came through from the lab I set about comparing it to the dozen or so Neanderthal sequences that have been published. It was very soon clear that Khwit's, and thus Zana's, mitochondrial DNA was not Neanderthal. This was a disappointment – it meant that Porchnev and Heuvelmans' beguiling theory was wrong. Zana was not a Neanderthal. However, the disappointment was strictly temporary. Zana may not have been a Neanderthal but when I compared her detailed mitochondrial DNA sequence with my database of hundreds of thousands of records from all over the world, there was a major surprise. The detailed sequence from Khwit's tooth showed beyond any doubt that his mitochondrial DNA, and therefore Zana's, was not from the Caucasus or anywhere close. It was from sub-Saharan Africa. I was stunned. Just like the genetic connection revealed in the last chapter between the Himalayas and the Arctic islands of Svalbard, here was DNA that was completely out of place. How could it have happened that Zana, living wild in the forests of the Caucasus, had DNA from thousands of miles away on another continent?

Once I had calmed down I began to think rationally. Though there was no doubt about the accuracy of this result, it did not necessarily follow that Zana herself was fully African. Many people in Europe and America, for example, have inherited African mitochondrial DNA from remote matrilineal ancestors even though the rest of their genomes are unremarkably indigenous. Zana might have had a distant matrilineal ancestor from Africa but the rest of her DNA could well have been regular Caucasian. I realised this was something I could discover through the saliva samples that Dmitri Pirkulov was busy collecting in Abkhazia. He had started collecting these to check on Zana's Neanderthal credentials before we had the tooth result.

As we have already covered, between 2% and 4% of the nuclear genomes of Europeans and Asians, though not Africans, has a Neanderthal origin. It is straightforward to estimate the Neanderthal component in any individual's genome, and several genetic testing companies offer this as an entertaining adjunct to their normal products. I had planned using Pirkulov's saliva samples to estimate the Neanderthal component in the genomes of Zana's descendants. When I realised I could also use the same data to estimate their African component, I emailed Pirkulov at once to see how he was getting on, and also to ask him to collect some unrelated Abkhazians to be the controls I knew would need.

Dmitri Pirkulov did a brilliant job in Abkhazia. He managed to collect saliva samples from all six of Zana's descendants and also from a dozen unrelated locals. As soon as I received them, I sent them straight off to the lab. Although I had originally intended using these samples solely to estimate the Neanderthal percentage in the DNA of Zana's descendants, I knew the same data could also be computed in a different way to reveal any African DNA in their genetic make-up.

It took a couple of weeks for the lab to process the saliva samples, and I was nervously waiting for the results. First, had they worked? Some of the collection tubes had leaked, others were very messy with particles of food swilling around in the precious fluid. It would be very hard to get repeat samples if these failed. Were they from the right people? I didn't doubt Dmitri Pirkulov's efforts at locating Zana's descendants, but even with the best of intentions I knew from previous experience with genealogy that family recollections can sometimes be shaky. Would they contain any African DNA? If they did, then this would confirm the African origin of the DNA from Khwit's tooth. If not, it would make the tooth results interesting but ambiguous. Only if there were obvious signs of African DNA in Zana's descendants would we know whether her African ancestry was substantial.

The lab results arrived as an email attachment in the early hours of a Saturday morning. I normally don't look at my emails at the weekend, but this was an exception. As dawn broke over the River Cherwell I downloaded the files. Despite the unpromising appearance of some of the saliva, I had results from all of them, which was a great relief. I immediately started computing the DNA results to tell me how much Neanderthal DNA there was in Zana's descendants and, crucially, if any of them carried African DNA.

The Neanderthal content of Zana's six descendants was unremarkable, varying over a narrow range of between 2.6% and

2.8%. The same was true of the unrelated Abkhazians that Dmitri Pirkulov had collected as experimental controls. Had Zana been a Neanderthal, the genomes of her descendants nuclear would have contained a far larger proportion of Neanderthal DNA than the average. However, already knowing from the analysis of Khwit's tooth that Zana was not a Neanderthal, it was no surprise that her descendants had no more Neanderthal DNA in their nuclear genomes than regular Abkhazis. Moreover, the Neanderthal components of both Zana's descendants and the local controls were well within range of the average European values and had clearly not been elevated by recent interbreeding, as I had thought they might have been. It was a long shot, but worth a go.

More importantly, had I not tried I would not have asked Dmitri Pirkulov to collect saliva samples from Zana's descendants in the first place and could not have carried out the next crucial step in the reconstruction of Zana's genetic identity. I switched the program to reveal the African component from the same data. By then it was morning. I ate some breakfast as much to slow me down as anything. I wasn't in the least hungry. I realised I might be on the edge of a major discovery. That hasn't happened very often during my career, but the thrill of being the very first person in the world to discover something has always been at the root of why I became a scientist. I paced myself.

I would soon be the first to know whether or not Zana's descendants had inherited any of her African DNA. Khwit's tooth had already told me that Zana's matrilineal ancestors had come from Africa, but not how much, if any, of the rest of her genome was also African. I managed to do things in the right order. Before looking at the results from Zana's descendants I fed in the data from the controls, the unrelated Abkhazis that Dmitri Pirkulov had sampled. The screen displayed each of their chromosomes in segments whose colour indicated their origin: blue for European, green for Asian and red for African. The first control showed blue and green segments but no red. The second, though having a different pattern of blue and green on their genetic portraits, none were red. So it continued. After half an hour I knew that all twelve of the unrelated Abkhazis had absolutely no African DNA in their genomes.

I reformatted the program and prepared to enter the data from Zana's six descendants. This was the moment of truth. I pressed the key that started the program and over an agonising two minutes the chromosome portrait of Zana's great-great-grandson, Zoya Makarian, materialised on the screen. Among the blue and green were seven long and eight short segments of DNA that were bright red. African! The program calculated that 5.9% of Zoya Makarian's genome was from Africa. I carried on to the next descendant, Manana Jologua, Zana's great-granddaughter. She had even more long stretches of red in her chromosome portrait, estimated by the program at 8.8% higher than Zoya Makarian's, probably because she was one generation closer to Zana. When I painted the chromosome portrait of Zana's great-great-great-granddaughter Indira Haytzuk she certainly had African segments, but only 2.7% of her genome was coloured red, a reflection that she was one generation further removed from Zana than Zoya Makarian and two more than Manana Jologua. By the time I had the genetic portraits of Zana's six descendants and found segments of African DNA in all of them I was exhausted, but I knew I had made a very important discovery.

I added the figures for their African component to the pedigree given to me by Dmitri Pirkulov, which is shown in an appendix. The African component of Zana's descendants ranged from 2.7% to 8.8%. One was a grandchild, four were Zana's great-grandchildren, and the sixth was a great-great-grandchild. If the African component in their genomes had come from Zana, which is a reasonable assumption given that the regular Abkhazi genomes that made up the rest of the ancestry had none, then I could calculate the African component in the DNA of Zana herself, even though she had died over 120 years ago. In each generation of her descendants, Zana's genetic contribution would be roughly halved. Even allowing for the expected random fluctuations, it was straightforward to show from these figures just how much of Zana's nuclear DNA was African. The details are confined to an appendix but the answer is very simple. Zana's DNA was 100% African. Stunning.

How was it that a full-blooded sub-Saharan African woman came to be living wild in the foothills of the Caucasus Mountains in the middle of the nineteenth century? None of the explanations for this remarkable fact are straightforward.

There had been a few African slaves in Abkhazia in earlier centuries when it was part of the Ottoman Empire and, theoretically at least, Zana could have been an escaped slave. The difficulty with this most prosaic of explanations is that although the slave theory might explain her African DNA it certainly does not account for her remarkable appearance. For a start, she was nothing like a modern African in her looks or her behaviour: 'Zana had all the characteristics of a wild animal . . . the most frightening feature was her expression which was pure animal, not human . . . She dug herself a hole in the ground and slept in it . . . she walked naked even in winter, tearing dresses that she was given into shreds . . . her athletic power was enormous. She would outrun a horse and swim across the Moskva River even when it rose in violent high tide.[1] Feral children and adults are rarely healthy, and are usually discovered on the verge of starvation, yet Zana possessed superhuman strength and athleticism. Is it likely that an escaped slave girl could have sustained herself in the wild, and so well that she developed her remarkable physical attributes? Almost certainly not.

The other contradiction was that Zana's mitochondrial DNA, deduced from the results from her son's tooth, had closer matches with West rather than East Africa. The Ottoman slaves were all from East Africa, usually coming through the slaving port of Zanzibar off the coast of Tanzania.

Zana's vigorously healthy physical condition bears all the hallmarks of a life in the forest lived not alone but in the company of others, as does the similarity between descriptions of her appearance and the numerous eyewitness accounts of *almasty* collected by Koffman from all over the Caucasus during her countless expeditions. These consistently describe an elusive forest people with dark skin covered in red-brown hair. Zana never spoke, not to herself, nor to her captors. Even a community of escaped slaves living wild in the forest would have conversed with one another and Zana would surely have made some attempt at doing so with her captors. Even if she did not know Abkhazi, there is always scope for some sort of verbal communication, and in time she would surely have picked up a few words of the language of her captors. But there was nothing. Not a word, not a gesture, not even the smallest attempt at communication.

The well-researched contemporary descriptions suggest to me that Zana had nothing to do with the modern world. Porchnev and Heuvelmans were quite reasonable in their hypothesis that Zana was a surviving Neanderthal but we now know from the mitochondrial DNA analysis of her son's tooth that she was not. But if not a Neanderthal, was she fully human? She was certainly a member of the genus *Homo*. That much was certain from the fact that she produced healthy, fertile children. We know from earlier chapters that hybridisation between different human species has certainly occurred in the past and so it is conceivable that her sons Eshba and Khwit were hybrids. Zana may have been in the genus *Homo* without being fully *Homo sapiens*.

Are there clues in the mitochondrial DNA sequence recovered from Khwit's tooth? The details place Zana in the clan of Lingaire, also called L2C, one of the oldest of the thirteen matrilineal African clans. All African clans are very ancient, much older than any outside Africa, for the simple reason that humans have been in Africa far longer than anywhere else. The clan of Lingaire is around 150,000 years old, ancient even for Africa.

Our *Homo sapiens* ancestors left Africa to settle in the rest of the world around 100,000 years ago. It was neither a large nor representative exodus, and the members of only one of the thirteen African clans took part. This was the clan of Lara, also known as L3A. Though Zana's mitochondrial DNA clearly establishes her ultimate origin as African it does not tell us when her ancestors left, though it was not typical of the main Laran exodus.

The clan of Lingaire is significantly older than the date of that diaspora, so Zana's ancestors could have left Africa *before* the Laran exodus of 100,000 years ago. Zana would then be a survivor from an African diaspora that fizzled out in the face of competition – the later spread of *Homo sapiens* that left her ancestors hanging on in the remote valleys of the Caucasus.

To begin to answer this intriguing question, I checked to see if there were any matches with Zana's mitochondrial sequence in any of the available databases. There were none that matched exactly. I also had the scraps of her nuclear genome scattered among her descendants. Again, the African segments did not match any records. Throughout *The Nature of the Beast* I have been critical of the speculation of others, and of premature disclosures. Suffice to say that the scraps of Zana's DNA that have floated down through time like fragments of a faded photograph, to her descendants are very, very unusual. I am hard at work making sense of it and I hope to know soon whether Zana was indeed a survivor of an antique race of humans.

If Zana's people were in the Caucasus during the nineteenth century when she was captured, they might well be still there to this day, living as they have for millennia somewhere in the wild valleys that radiate from the eternal snows of Elbrus.

30

Finale

In the end there was no need to trouble David Hume. Only if the DNA results from Bigfoot or yeti hairs had proven to be anything other than ordinary animals would I have needed to approach the great philosopher on his pedestal.

The exception was the yeti mummy that Christophe Hagenmuller brought back from Ladakh. That may have been the 'different kind of bear' that Reinhold Messner had consistently believed was the basis of the yeti legend. Perhaps it was more aggressive, retaining some of the habits of its polar bear ancestor, the only carnivore that regularly hunts and kills humans. But there was no need to consult Hume about this. Not yet anyway. The road ahead was clear. We need more DNA sequence information on this yeti-bear and that is already on its way. It might indeed be a 'different kind of bear' but it was not a primate, or a surviving Neanderthal or any other kind of hominid.

Neither were any of the Bigfoot samples from America. They

had all proven to be just regular animals. However convinced the donors were of their authenticity, they had not come from any primate sasquatch, but from very ordinary mammals. Horse and cow were high on the list, both in the US and in Russia, and that may well be because so many of the samples came from the sites of Bigfoot or *almasty* encounters of one sort or another, rather than from the creatures themselves. Horses and cows have long, thick hairs that are more easily seen caught on bushes close to a 'hotspot', as many of these hairs had been. Clearly these samples of feral domesticates are straightforward misattributions. Only bears could be realistically confused with a yeti or a Bigfoot under favourable conditions, even if the good people of Teslin, Yukon, mistook the backside of a bison for a sasquatch.

The most likely explanation is that the donors, all enthusiasts, got carried away and didn't pause to consider that a hair on a bush close to where they saw or heard a Bigfoot is not certain to have come from the creature. Sightings too might not be what they seem. The phenomenon of pareidolia describes the ability of the mind to form images that are not really there, like 'the man in the moon' or the famous case of the face of the Virgin Mary on a piece of burnt toast. Pareidolia has probably evolved as a warning mechanism with survival advantages in the past. Michal Heaney, who was kind enough to direct me to Jeanne-Marie Koffman's original work on the *almasty*, put it very succinctly: 'Far better to mistake a tree for a leopard, than a leopard for a tree.' Pareidolia could account for some misattributed sightings, though it may be impertinent to suggest such a thing.

It has never been my purpose to explain what so many witnesses have seen or heard or indeed smelled. I set out with as open a mind as possible to test the organic remains that have been attributed to these creatures and to identify what species they belonged to. I would not have spent the effort unless I thought there was a realistic possibility, small though it may have been, of finding an anomalous primate, even if

most samples were likely to be something more mundane, as
turned out to be the case. There was one time, as you'll have
seen in Chapter 25, when I was on the brink of losing my own
scientific detachment and was only saved by the admirable
logic of Sage, the park ranger. Under those conditions of high
anxiety I would have been convinced that any hair I found
was from a sasquatch.

From what I have seen, Bigfootologists are not, on the whole,
good researchers. They lack the necessary degree of self-criti-
cism. One of the elements of scientific training is that you
should be your own fiercest critic, though many of us fail to
live up to this dictum. You don't have to be right every time
– indeed progress in science is a process of evolution where
one theory supersedes the last, however strongly held, as new
information or new thinking is revealed. All of it is based on
evidence, testable, repeatable and – most important – publish-
able in peer-reviewed journals[1]. There has been precious little
of that in the search for anomalous primates. I hope, however,
that I have shown that despite its appalling record the search
for yeti, *almasty* and Bigfoot is not beyond the scope of science.
I may not have found an anomalous primate among the hair
samples I was given, but that is simply because there wasn't
one, nothing more.

This project certainly could never disprove the existence of
the yeti and other similar creatures, and that was never its purpose.
On the contrary, it has shown that a single hair is enough to
make an unambiguous identification. The next one might always
be the 'golden hair' that provides the final proof. Many of my
donors, having learned that the hair they thought was from a
Bigfoot was actually from a bear or some other animal, have
returned to the forests encouraged that there is now at least a
way of proving and identifying what they know in their hearts
is still out there.

Throughout the book you have heard the stories of the aficio-
nados and the extraordinary energy with which they follow their
passion. I am quite sure that will continue, and I hope it does,

with perhaps more direction to their search for proof than previously. There is no reason to abandon the search just because none of the hairs donated to this project was from an anomalous primate. There will be better ways of obtaining hair samples, better traps laid on known Bigfoot trails. Improve the way you research. Try to innovate. Already Rhettman Mullis is planning a programme of catch and release. I hope he succeeds.

Money is an issue. If the first two big questions about Bigfoot and the yeti are, 'Do they exist?' and, 'What are they?' the third is surely, 'Who's paying?'. These tests are not cheap, at around a thousand pounds for each mitochondrial DNA analysis, and lie beyond the budget of most Bigfoot enthusiasts. I have pointed out what happens when samples are sent to laboratories without the funds to pay for the analysis. I certainly hope for an improvement here, both from Bigfootologists and any scientist who accepts a sample for analysis. For the enthusiast, choose a lab that has a good record of scientific publication and knows how to handle contaminated samples in poor condition. Insist on a proper report, not a throwaway opinion, especially of the 'somewhere between an ape and a human' variety. You will have wasted a potentially precious sample for nothing. For scientists, who I do feel I am in a position to criticise, just say no to samples you aren't going to treat properly. If you do accept a sample, then treat its analysis with the respect its donor is entitled to expect.

Given its popularity for the 'gazing populace', there is always going to be scope for hoaxers. But even that might change. After a recent Bigfoot 'shooting' claimed by the notorious hoaxer Rick Dyer, he was set to go on the lucrative lecture circuit through a well-known agency with an international portfolio of celebrated speakers. I was asked if I would run a DNA test on Dyer's Bigfoot body, which I agreed to do, only to find that Dyer – whose website asks the question 'Why won't anyone f...ing believe me?' – declined this opportunity to verify his exhibit's authenticity. It is reminiscent of the case of the Minnesota Iceman; the main difference being that while the Iceman was a professional hoax,

Dyer's creature, currently on display in a parking lot in Las Vegas, is a pale imitation in a plywood coffin and not even frozen. Now that there is a way to verify these claims, how long will it be before the public tires of being dished up such obvious nonsense, the ratings fall away and we are spared any more of this mischievous and mendacious baloney? Not soon enough in my view.

Two unexpected and promising leads have come from this project. There may be a 'different type of bear' wandering the Himalayas, and there are expeditions leaving soon to find one alive. The case of Zana, the wildwoman from the Caucasus, is also a very rewarding outcome. We now know a lot more about this famous case, thanks mainly to the persistence of Igor Burtsev and his colleagues from Moscow's Darwin Museum. Their diligence and effort in finding and sampling Zana's relatives is a great achievement. With further genetic investigations actively under way, they might soon be able to say they have found perhaps not a Neanderthal survivor, but an antique race of humans living in the Caucasus.

I have often been asked whether I believe the yeti exists. Up to now I have refused to answer, lest it stops me having the open mind I needed. It was also an irrelevant question since I was trying to find some evidence on which to base an opinion. Funnily enough, even though there were no anomalous primates among the hairs I tested, I think my view has altered more in favour of there being 'something out there' than the reverse. This change of heart comes from speaking to several people, some not even mentioned in *The Nature of the Beast*, who have nothing to gain but who have seen things, in good light while in the company of other witnesses, that are hard to explain otherwise. To automatically reject these accounts is just as blinkered as accepting that every broken branch has been snapped or twisted by a sasquatch.

One day soon I hope to be able to approach David Hume on his pedestal with DNA evidence of anomalous primates, either my own or someone else's, that even he will accept. But spare a last thought for the yeti, Bigfoot and sasquatch peering unseen from behind a tree at those who would unmask her in the name of progress. If such creatures do exist, one thing is certain. They want nothing to do with us.

Notes

2: The Yeti Enigma

1. Rawicz, S. 1956. *The Long Walk*. Constable, Edinburgh.
2. Chase, S. quoted at: http://www.yowiehunters.net.

3: The Last Neanderthal

1. Greig. D.M, 1933, A Neanderthaloid Skull Presenting Features of Cleidocranial Dysostosis and Other Peculiarities, *Edinburgh Medical Journal* **40**, 497.
2. Fleure, H.J. 1951. *The Natural History of Man in Britain*. Collins, London.
3. Howells, E. 2005. *Good Men and True: The Lives and Tales of the Shepherds of mid-Wales*. Capel Madog, Aberystwyth.

4: The Footprint that Shook the World

1. Hillary, E. and Doig, D. in Napier, J. 1972. *Bigfoot*. p.133, Jonathan Cape, London.

5: The Professor

1. Ward, M. 1997. *Wilderness and Environmental Medicine* **8**, 29–32.

7: The Russian *Almasty*

1. Regal, B. 2013. *Searching for Sasquatch: Crackpots, Eggheads and Cryptozoology*. Palgrave MacMillan, London.
2. Halbertsma, T. 2011. 'Mongolia's "Homo sapiens Almas"', *Kraken* **2**, 41–57.

8: The Godfather

1. Heuvelmans, B. 1969. *Bull. Inst. Sci. Nat. Belg* **45**, 7–24.

9: Clutching at Straws

1. Coleman, L. 1989. *Tom Slick and the Search for the Yeti*, p.87, Faber & Faber, Boston.

10: Our Human Ancestors

1. Green, R. et al. 2010. *Science* **328**, 710–22.
2. Abi-Rached, L. et al. *Science* **334**, 89–94.
3. Pääbo S. 2014. *Neanderthal Man: In Search of Lost Genomes*. Basic Books, New York.
4. Krause, J. et al. 2010. *Nature* **464**, 894–7.
5. Pääbo S. *ibid*. p.235.
6. Hammer, M. et al. 2011. *Proceedings of the National Academy of Science of the United States of America* **108**, 15123–8.
7. Meyer, M. et al. 2014. *Nature* **505**, 403–6.

11: Keeping it in the Family

1. Van Gelder, R. 1977. *Novitates* no.2365, 1–25, American Museum of Natural History, New York.
2. Bernolet-Moens, H. 1908. Privately published pamphlet, Heuvelmans Archive, Museum of Zoology, Lausanne.
3. Chilvers, H. 1930. *The Seven Lost Trails of Africa*. p.7. Cassell & Co, London.
4. Etkind, A. 2009. *Studies in the History and Philosophy of Science* **39**, 205–210.
5. Heuvelmans, B. and Porchnev, B. 2011, *L' Homme de Néanderthal est toujours vivant*. Les editions de l'oeil du Sphinx, Paris.

14: Good Science, Bad Science

1. Meldrum, J. and Schaller, G.B. 2006. *Sasquatch: Legend Meets Science*, pp.266–7. Tom Docherty Associates, New York.
2. Milinkovitch, M.C., Caccone, A. & Amato, G. 2004. *Molecular Phyogenetics and Evolution* **31**, 1–3.
3. Matthiessen, P. 1998. *The Snow Leopard*, pp.75–80. Vintage Books, London.
4. http://www.lanevol.org/LANE/yeti_3.html.
5. Coltman, D. and Davis, C. 2006. *Trends in Ecology and Evolution* **21**, 60–1.

26: The Russians

1. Daily Mail Online. 12 February 2013 http//www.dailymail.

co.uk/video/news/. . ./Is-proof-Yeti-sighting-Siberia.html

27: The Laboratory Reports

1. Fahrenbach, H. 1997. *Cryptozoology* **13**, 47–75.
2. Meldrum, J. and Schaller, G.B. 2006. *ibid.*
3. The control region sequence of sample #25213 varied from the reference sequence at positions 16,234, 16,311 and 16,346, the last position being heteroplasmic.

28: The Snow Bear

1. Smythe, F.S. 1936. *The Valley of Flowers.* p.144. Hodder & Stoughton, London.
2. Brunner, B. 2007. *Bears: A Brief History.* p.64, Yale University Press.
3. Lindqvist, C. et al. 2010. *Proceedings of the National Academy of Science of the United States of America* **107** (11), 5053–7.
4. Talbot, S. and Shields, G. 1996. *Molecular Phylogenetics and Evolution* **5**, 567–75.
5. To complicate matters, an even closer DNA sequence match has recently come to light. This is with modern polar bear living on the Diomedes Islands in the Bering Strait.

29: Zana

1. Bayanov, D. 1996. *In the Footsteps of the Russian Snowman,* pp.46–52, Crypto-Logos, Moscow.
2. Heuvelmans, B. and Porchnev, B. 2011. *ibid.*
3. Koffman, J.-M. 1991. 'L'Almasty, yeti du Caucase', *Archeologia* **269**, 24–43.

30: Finale

1. I am delighted to say that a paper entitled 'Genetic analysis of hair samples attributed to yeti, bigfoot and other anomalous primates' was published in the July 7th 2014 volume of the prestigious journal Proceedings of the Royal Society. As far as I am aware this is the first occasion when a full paper on the topic has been published in a peer-reviewed science journal. It is available at: http://rspb.royalsocietypublishing.org/content/281/1789/20140161.full?sid=3e137e23-4ace-4ba1-a7b9-260630368149

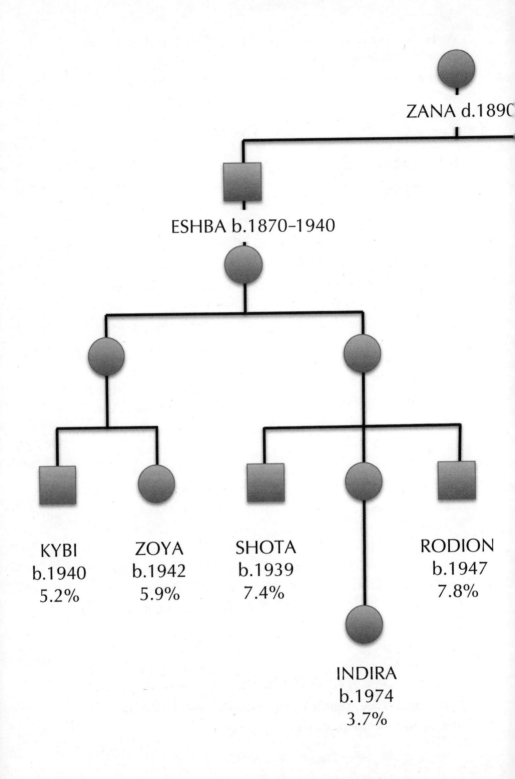

ZANA d.1890

ESHBA b.1870–1940

KYBI
b.1940
5.2%

ZOYA
b.1942
5.9%

SHOTA
b.1939
7.4%

RODION
b.1947
7.8%

INDIRA
b.1974
3.7%

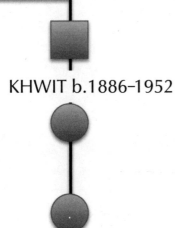

KHWIT b.1886–1952

MANANA
b.1963
8.8%

ZANA and her DESCENDANTS
Showing percentage of African DNA

Acknowledgements

Researching *The Nature of the Beast* has been a long journey from a crazy idea to a finished book, a scientific publication and a television documentary so there are a lot of people to thank.

In chronological order, I want to thank those of my colleagues in Oxford, especially Sir Anthony Epstein FRS who encouraged me that this was a perfectly sound scientific project. Without similar encouragement from Michel Sartori and Ken Goddard it would have been a lonely struggle to complete it. Michel, of course, deserves extra thanks for lending the name of his museum to the enterprise as a collaborator while Ken made me very welcome in the US Fish and Wildlife Forensic Laboratory in Ashland, Oregon.

I have met many wonderful people as I gathered the all important material on which this book is based. Some, like Reinhold Messner, are celebrities on the world stage while others, like Peter

Byrne, Adam Davies, Jonathan Downes and Rhettman Mullis are the celebrities of the cryptozoology community. I am also indebted to Rhettman for introducing me to so many of his contacts and travelling with me to remote locations in America to collect samples.

Which brings me to the many people who donated samples for me to analyse. They were brave to do so, knowing that genetics is no respecter of personal opinion, however deeply felt. So I am very grateful to Justin Smeja, Bart Cutino, Derek Randles, Dan Shirley, Garland Fields, Loren Coleman, Betty Klopp, Marcel Cagey, Sam Cagey, Lori Simmons, Mike Long, Patrick Spell, Maxwell David, Mark McClurkan, Rob Kryder, Jack Barnes, Jeff Anderson, David Ellis, Steve Mattice, Brenda Harris, Stuart Fleming, Igor Burtsev, Dmitri Pirkulov, Michael Trachtengerts and Dmitri Bayanov for submitting samples and for their progressive stance in doing so. Thanks also to Peter Byrne, Ray Crowe, Ronnie Coleman, Greg Roberts and Tom Graham for discussing their experiences and to Jeff Meldrum and Anna Nekaris for their advice and guidance.

As well as Ken Goddard I would like to thank Ed Espinoza, Mike Tucker, Barry Baker, Bonnie Yates, Cookie Smith and Dyan Straughan of the US Fish and Wildlife Service Forensic Laboratory, Ashland, Oregon for help with hair identification and introducing me to forensic methods of trace evidence analysis.

Terry Melton has been the rock on which the entire project has rested. Thanks to her skills, and those of her colleagues Charity Holland, Bonnie Higgins, Gloria Dimick, and Michele Yon, I was able totally to rely on the results from her laboratory.

Making 'The Bigfoot Files' was great fun and my thanks go to Harry Marshall, the creative director of Icon Films for deciding to follow the project well before we had any results and to Sara Ramsden from Channel 4 for having the courage to commission the film. Thanks also to Brendan McGinty, Sam Challenger, Claire

Efergan, Kate Edwards, Mark Evans and Steven Clarke, the people who made the film, for welcoming an ingénue on the set. I want also to thank Paul Smith, Director of the University Museum of Natural History in Oxford for allowing me to pluck some hairs from Mandy, the Shetland pony and other exhibits as controls and also for letting me be filmed in his magnificent office in the museum as if it were my own.

My agent, Luigi Bonomi, has kept me going, as always, with his infectious enthusiasm while Emily Hayward helped me through the unfamiliar maze of film contracts. Both deserve my thanks, as do my researcher Marcus Morris who coaxed the best from interviewees he met on my behalf, and my secretary Hilary Prince for keeping everything in order while I was away. Sue Foden's reading of my early drafts helped to get my manuscript into decent order. I was very fortunate to have Mark Booth as my editor at Hodder. His gentle yet firm suggestions improved my text beyond measure, as did Tara Gladden's eagle-eyed copy-editing, while Fiona Rose kept everything flowing.

Closer to home, my son Richard greatly enhanced *The Nature of the Beast* with his fine sketches at the beginning of each chapter, while my wife Ulla came with me when she could and, as always, sprinkled stardust wherever she went.

An invitation from the publisher

Join us at www.hodder.co.uk, or follow us
on Twitter @hodderbooks to be a part of
our community of people who love the very
best in books and reading.

Whether you want to discover more about a book
or an author, watch trailers and interviews, have the
chance to win early limited editions, or simply browse
our expert readers' selection of the very best books,
we think you'll find what you're looking for.

And if you don't, that's the place to tell us what's missing.

We love what we do, and we'd love you to be a part of it.

www.hodder.co.uk

 @hodderbooks

HodderBooks

HodderBooks